Lab Manual

biology

SIXTH EDITION

bju press®

Greenville, South Carolina

The writers and publisher have made every effort to ensure that the laboratory exercises in this publication are safe when conducted according to the instructions provided. We assume no responsibility for any injury or damage caused or sustained while performing activities in this book. Conventional and homeschool teachers, parents, and guardians should closely supervise students who perform the exercises in this manual. More specific safety information is contained in the BIOLOGY Teacher Lab Manual Sixth Edition, published by BJU Press. Therefore, it is highly recommended that the Teacher Lab Manual be used in conjunction with this manual.

NOTE: The fact that materials produced by other publishers may be referred to in this volume does not constitute an endorsement of the content or theological position of materials produced by such publishers. Any references and ancillary materials are listed as an aid to the student or the teacher and in an attempt to maintain the accepted academic standards of the publishing industry.

BIOLOGY Lab Manual
Sixth Edition

Writers
David M. Quigley, MEd
Christopher D. Coyle

Biblical Worldview
Bryan Smith, PhD
Tyler Trometer, MDiv

Academic Integrity
Jeff Heath, EdD

Instructional Design
Rachel Santopietro, MEd
Danny S. Wright, DMin

Editor
Rick Vasso, MDiv

Cover Designer
Sarah Lompe

Designer
Rachel Nichols

Illustrators
Giulia Borsi c/o Lemonade
    Illustration Agency
Rachel Nichols

Production Designer
Maribeth Hayes

DesignOps Coordinator
Lesley Ramesh

Permissions
Maria Andersen
Sharon Belknap
Sarah Gundlach
Stacy Stone
Kathleen Thompson
Carrie Zuehlke

Project Coordinators
Heather Chisholm
Chris Daniels
Gary Santiago

Postproduction Liaison
Peggy Hargis

Photo credits appear on pages 300–301.

The text for this book is set in Adobe Gothic, Adobe Minion Pro, Adobe Myriad Pro, Bungee by David Jonathan Ross, FF Uberhand by Jens Kutilek, Free 3 of 9 by Matthew Welch, LiebeRuth by Ulrike Rausch, Raleway by The League of Moveable Type, Sketchnote by Mike Rohde, and Times New Roman PSMT.

All trademarks are the registered and unregistered marks of their respective owners. BJU Press is in no way affiliated with these companies. No rights are granted by BJU Press to use such marks, whether by implication, estoppel, or otherwise.

The cover photo is a close-up of the feathers of a green-winged macaw.

© 2024 BJU Press
Greenville, South Carolina 29609
Fifth Edition © 2017 BJU Press
First Edition © 1980 BJU Press

Printed in the United States of America

ISBN 978-1-64626-116-1

15 14 13 12 11 10 9 8 7 6 5 4 3 2 1

CONTENTS

CONTENTS

An Unexpected Discovery

"One sometimes finds what one is not looking for."

Alexander Fleming in his laboratory in London during World War II

A *Penicillium* culture

When Alexander Fleming woke up on September 28, 1928, he wasn't intending to revolutionize medicine, but, as he would later conclude, "I suppose that was exactly what I did." The funny thing is that it was a complete accident.

Alexander Fleming was a Scottish biologist with a reputation for being a brilliant scientist, though he let his laboratory get, well, a bit messy. He returned there after a family vacation and was checking on some petri dishes holding growing bacteria. One culture looked different from the others. An airborne fungus had evidently traveled to the dish and had begun to grow in that culture. It had killed the bacteria growing around it. "That's funny," Fleming remarked. The fungus was *Penicillium*, and this unexpected discovery was the beginning of a new era in medicine. Penicillin, the antibiotic derived from *Penicillium*, would save thousands of people from diseases like scarlet fever, diphtheria, and pneumonia. Fleming received the 1945 Nobel Prize for his discovery.

Though you won't be playing with dangerous bacteria this year in biology, you can develop skills to help you work like a biologist at your own level. As you work through this lab manual, occasionally you'll have to think about how to solve a problem on your own without any procedures spelled out in the lab activity. These activities will have the word "inquiring" in the subtitle. You'll learn how to obtain useful data from the tools you have to work with, things as different as smartphone apps, microscopes, and good old-fashioned glassware.

This manual is your guide as you learn to use the tools and think the way a biologist does. Be sure to check the appendixes to learn about safety rules, equipment techniques, and writing formal lab reports. Read through the procedures and follow them. Answer questions interspersed with the procedures. Measure carefully. Ask if you aren't sure what to do; check your textbook for more information. Record data carefully in the tables, drawing areas, and graphing areas provided at the end (usually) of a lab activity. And keep your mind engaged! You never know what you may discover when you do.

But what good is all of this? Why is it important to develop the skills and mindset of a biologist? If you are a Christian, you're not just a student or even a student biologist. A Christian should see science as an amazing tool to glorify God and help people by obeying God's command to wisely use His creation. We should do the work of biology within the context of a Christian worldview and use it as God intended in ways that harmonize with His Word.

So ... let's get into the laboratory!

Safety Icons

Pay close attention to these icons whenever you see them.

 Animal

Animals that you are to observe or collect may inflict stings or bites.

 Body Protection

Chemicals, stains, or other materials could damage your skin or clothing. Wear a laboratory apron or laboratory coat and gloves as directed by your teacher.

 Chemical Fumes

Chemical fumes may present a danger. Use a chemical fume hood or make sure that the area is well ventilated.

 Electricity

An electrical device (hot plate, lamp, microscope) will be used. Use the device with care and watch for any frayed cords.

 Extreme Temperature

Extremely hot or cold temperatures may cause skin damage. Use proper tools to handle laboratory equipment.

 Eye Protection

There is a possible danger to the eyes from chemicals or other materials. Wear safety goggles.

 Fire Hazard

A heat source or open flame is to be used. Be careful to avoid skin burns and the ignition of combustible materials.

 Pathogen

Organisms encountered in the investigation could cause human disease.

 Plant

Plants that you are to observe or collect may have sharp thorns or spines or may cause contact dermatitis (inflammation of the skin).

 Poison

A substance in the investigation could be poisonous if ingested.

 Sharp Object

When using equipment in this lab activity, be careful to avoid cuts from sharp instruments or broken glassware.

1A LAB

A Method to This Madness

Scientific Inquiry

Laura wants to use scientific inquiry to test whether practice can improve reaction times. To be sure that her results are valid, she decides to conduct a controlled experiment. A controlled experiment typically has two groups that are identical except for a single factor, called the *experimental*, or *independent*, *variable*. The group exposed to the independent variable is called the *experimental group*. The second group, the *control group*, is not exposed to the independent variable. During the experiment, the researcher measures a factor in both groups. This factor is called the *dependent variable*—it results from, or is dependent on, the independent variable. The experiment is designed to determine whether there will be a measurable difference in the dependent variable between the two groups.

Six students at Laura's school agree to be part of her experiment. To consistently measure their reaction times, she decides to use a falling meter stick. The test subject will have to catch the falling meter stick after seeing it begin to fall. The distance that it falls before being stopped will indirectly measure reaction time. She formulates a hypothesis: The reaction time of students who have practiced catching the meter stick will be less than their reaction time before having practiced.

Each of the six students places a thumb and forefinger on the edge of a ring stand and keeps eyes closed. Laura positions the meter stick in the center of the ring stand so that the end is level with the student's fingers. At Laura's instruction, the student opens his or her eyes and Laura drops the meter stick. Her friend records how far the meter stick drops before the student catches it. Laura repeats these procedures with each student and averages the results.

Laura repeats the experiment five more times and plots the six averages on a bar graph to see whether student reaction times have improved with practice. The independent variable is the amount of practice, and the dependent variable is the reaction time.

How do scientists find answers to their questions?

Equipment

meter stick

PROCEDURE

Now it's your turn. You will conduct a different experiment with a different experimental variable. You will test whether a person can catch a meter stick faster if given a heads-up that it is about to be dropped. Your classmates will form both the experimental and control groups.

QUESTIONS

- How can you use scientific inquiry to answer questions?
- What is a controlled experiment?
- How is a controlled experiment used to answer scientific questions?

1. Scientists often start out with a research problem—the question they are trying to answer. What is your research problem?

_____

_____

_____

_____

2. What is your hypothesis?

_____

_____

_____

_____

Your teacher will be the "caller" and will divide the class into teams consisting of a "dropper" and a "catcher." Each team should record its own results. Remember, each lab group is part of two groups: the experimental group and the control group.

3. What are three variables that you need to keep constant?

_____

_____

_____

4. What is the independent variable?

_____

_____

5. What is the dependent variable?

_____

_____

After each pair of droppers and catchers is provided with a meter stick, the droppers will hold it up for the catchers to catch, keeping constant the variables identified in Question 3.

A *Experimental Group:* The caller calls out a sequence such as "1, 2, 3, drop," or "Ready, set, go." This will alert droppers to drop their meter sticks in unison. Record the distance that the meter stick fell before it was caught in the *Trial 1* row of Table 1.

B *Control Group:* The experiment will now be repeated, but the caller will stand where not visible to the catchers and will signal silently to the droppers to drop their meter sticks. Record the distance the meter stick fell in the *Trial 1* row of Table 1.

C Repeat Steps A and B four times, recording the results for Trials 2–5 in Table 1.

ANALYSIS

D For each team, average your experimental group scores and your control group scores and record the averages in Table 1. Make a bar graph of the data in Graphing Area A.

E Copy the averages from Table 1 to the first row of Table 2. Obtain the experimental group averages and the control group averages from your classmates and record them in Table 2. Average these averages and record the results in Table 2. Create a bar graph of the data in Graphing Area B.

6. What conclusions can be drawn from the data?

_____

_____

_____

_____

7. Does the data support the hypothesis? Explain.

_____

_____

_____

_____

_____

PERSONAL OBSERVATIONS

8. Did your repeated trials yield similar results? If not, suggest some sources of error that might have affected your results.

_____

_____

_____

_____

_____

9. What could you do to make the repeated trials of your experiment closer in value?

_____

_____

_____

_____

10. Can you think of any additional changes that need to be made?

_____

_____

_____

_____

_____

_____

11. On the basis of your experience, what other experiments dealing with response time would you like to try?

_____

_____

_____

_____

GOING FURTHER

You read a news article about a scientific study into whether Kentucky bluegrass (*Poa pratensis*) or timothy (*Phleum pratense*) is a better grass for raising horses. A herd of thirty thoroughbreds was pastured in a field of Kentucky bluegrass in Tennessee, and a herd of thirty mustangs was pastured in a field of timothy in Arizona. At the end of the yearlong study the average weight gain of the thoroughbreds was greater than the average weight gain of the mustangs. The study concluded that Kentucky bluegrass is a better fodder for horses.

12. This study has some problems, primarily with keeping the two groups the same except for the experimental variable. Redesign this experiment in a way that corrects these problems.

_____

_____

_____

_____

_____

_____

_____

_____

_____

TABLE 1 Group Data

	Experimental Group Distance (cm)	Control Group Distance (cm)
Trial 1		
Trial 2		
Trial 3		
Trial 4		
Trial 5		
Average		

TABLE 2 Class Data (Averages)

	Experimental Group Distance (cm)	Control Group Distance (cm)
Your Averages		
Classmate's Averages		
Classmate's Averages		
Classmate's Averages		
Classmate's Averages		
Average		

GRAPHING AREA A

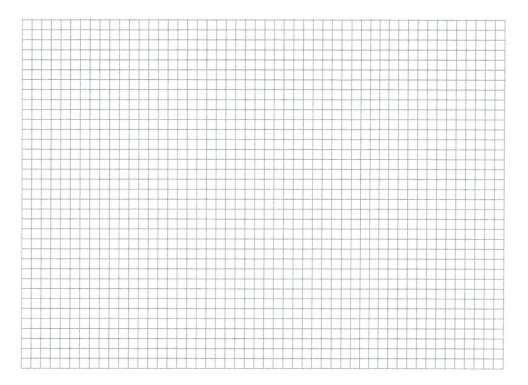

GRAPHING AREA B

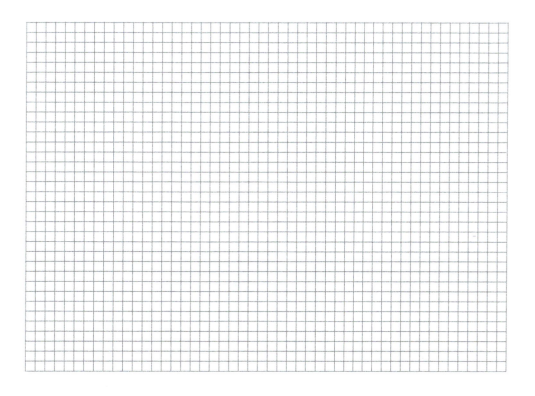

1B LAB

More Than Meets the Eye

The Microscope

In 1676 a Dutch cloth merchant named Antonie van Leeuwenhoek was investigating what gives pepper its hot taste. His hobby was making lenses, and with them he had already discovered many things too tiny to see with the unaided eye. But on this particular day he saw creatures much smaller than he or anyone else had ever seen before: bacteria.

Van Leeuwenhoek made and used simple (single-lens) microscopes because the low-quality compound microscopes of his day distorted images. His simple microscopes could sometimes render clear images at magnifications four times greater than existing compound microscopes. The secret to van Leeuwenhoek's success was his skill at making lenses, using a technique that he kept secret. But by the middle of the 1800s lens designers had greatly improved the compound microscope, allowing microbiologists to see things with greater clarity. Today almost all light microscopes are compound microscopes; simple microscopes are usually called *hand lenses* or *magnifying glasses*. A compound microscope is a very important tool in almost any field of biology.

In this lab activity you will learn how to use a microscope to study tiny life-forms, much as van Leeuwenhoek did almost 350 years ago.

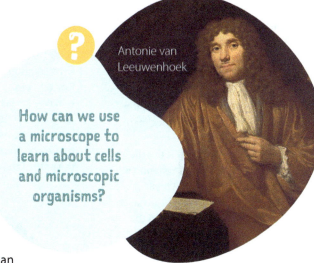

?

Antonie van Leeuwenhoek

How can we use a microscope to learn about cells and microscopic organisms?

QUESTIONS

- What are the parts of a microscope?

- What is the proper care for a microscope?

- How is a microscope used?

Equipment

microscope
lens paper
tissue

preserved slide of desmids or diatoms
immersion oil

PROCEDURE

The Structure of Your Microscope

As you study the infographic below, find each part on your microscope and check the box that goes with that part.

Adjustment Knobs ☐
Most microscopes have two types of adjustment knobs, one of each on each side of the microscope. The larger coarse adjustment knob allows you to rapidly change the distance between the specimen and the objective. The smaller fine adjustment knob is usually found underneath or centered on the coarse adjustment knob. It allows you to change the distance between the specimen and the objective slightly, producing a sharper focus.

Arm ☐
This "backbone" of the microscope supports the body tube.

Body Tube ☐
For a compound microscope to function properly, the two lenses must remain a certain distance apart. This long, narrow tube maintains that distance. In microscopes with an inclined body tube, mirrors are used to bend the path of the image.

Eyepiece ☐
This holds the top lens of the two lenses in a compound microscope. It is sometimes called the *ocular lens*.

Nosepiece ☐
This rotating disc at the bottom of the body tube holds the objectives, the metal cylinders that contain the bottom lenses of a compound microscope. Most microscopes have several different objectives that you can interchange by rotating the nosepiece. Each objective has a different power of magnification.

Stage ☐
This platform positioned directly below the objectives and above the light source supports the specimen.

Diaphragm and Substage Condenser ☐
These are located between the stage and light source. The diaphragm regulates the amount of light that passes through the specimen. The substage condenser provides a clearer image by bending and concentrating light before it reaches the specimen.

Stage Clips ☐
These fasteners on top of the stage hold the slide containing the specimen firmly in place.

Base ☐
This large rectangular or horseshoe-shaped structure supports the microscope and keeps it steady.

Light Source ☐
This is typically some type of light bulb, and in many microscopes you can adjust the amount of light being produced.

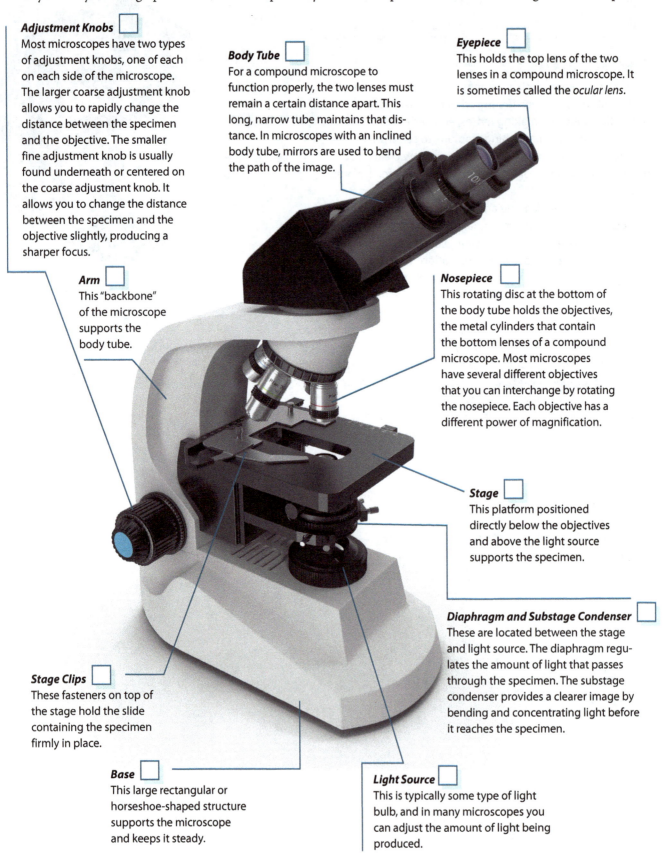

1. Which part of the microscope does the specimen rest on?

   _____

2. Which two structures hold lenses?

   _____

Caring for a Microscope

It is important to carry the microscope properly. Excessive jarring or bumping may knock the lenses out of adjustment.

A Carry the microscope with two hands—one under the base and the other on the arm.

B Keep the microscope close to your body in an upright position so that the ocular lens does not slip out of the body tube.

C Place the microscope gently on the table away from the edge. Remove the dust cover, if present, and plug in the power cord. Leave the light source turned off for now. Your microscope should have been stored with the stage in its lowest position and its nosepiece rotated to the lowest-power objective (typically 4×). Check to see that this is the case and make adjustments if necessary.

You will have a better experience using the microscope if you prepare it properly. Your microscope may need to be cleaned before you begin to use it.

D Use lens paper to clean lens surfaces and the light source.

3. Why should you not use a tissue to clean a lens surface?

   _____

   _____

E Wipe the lens in one direction across the diameter of the lens. Do not use a circular motion—doing so might grind dust into the lens, scratching it.

F Consult your teacher if any material remains on your objectives. Never use your fingernail or another object to chip away hardened material.

G Never attempt to take your microscope apart.

H Check that there is no slide on the stage.

Using Your Microscope

To have a clear view of your specimen, follow these procedures exactly. Keep in mind that it will probably take some practice for you to become proficient in using a microscope, so don't get discouraged if your first attempt at viewing a specimen doesn't give you the results you expect. With practice, using a microscope will become almost second nature, and you will be able to get a clear image quickly.

I Make sure that the nosepiece is rotated back to the scanning objective and that the stage is still lowered.

J When using your microscope, it helps to know the powers available. Power is the number of times larger the image of the object appears. You will find the powers written on the eyepiece and the objectives. As the power increases, the size of the image increases, but the field of view shrinks. Record the power of each objective in the appropriate rows of Table 1. Then compute, as directed in Table 1, the total magnification obtained using each objective and record the results.

K Obtain a preserved slide. Make sure that the slide is dry and free of cracks or other damage.

L Place the slide on the stage with the cover slip up, directly over the opening in the stage, and secure it with the stage clips. Turn on the light source.

M Position the slide so that the specimen is centered over the opening of the stage.

Now that you have the slide in place, you can focus your microscope.

N Double-check that the scanning objective (4×) is directly below the body tube. You should be able to feel that it is clicked into place.

O Looking at your microscope from the side, turn the coarse adjustment knob until the objective is just above the slide. (Some microscopes have safety devices that will prevent the objective from striking the slide and damaging the lens.)

P Look into the ocular lens and slowly turn the coarse adjustment knob until the specimen comes into focus. Make sure that the distance between the objective and the stage is increasing. You do not want the objective to strike the slide, potentially cracking or even breaking the slide.

Q Once the specimen has been brought into focus, slowly turn the fine adjustment knob to get a sharper image.

4. If you move the slide to the right, how does its position change in your field of view?

_____

_____

5. Turn the fine adjustment knob slowly. Describe what happens to the image that you are viewing. Why do you think this occurs?

_____

_____

_____

_____

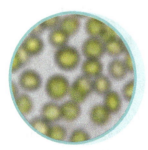 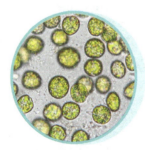

A slightly out-of-focus image (left) and a properly focused image (right)

R Draw the outlines of three specimens in Drawing Area A using the instructions provided in Appendix D; ignore the internal structures. Make sure that you are drawing typical specimens, not just odd globs that may be on your slide. Confirm the correctness of your drawings with your teacher.

Using High Power on Your Microscope

Now that you've learned to focus your microscope, you can switch to the next higher power.

S Center the desmid or diatom that you want to examine in the microscope field.

6. Why is it essential to position the specimen in the center of the field?

_____

_____

_____

_____

_____

T Focus your microscope using the scanning objective (4×) and then rotate the nosepiece to the low power objective (10×). If necessary, focus the image using the fine adjustment knob.

U Rotate the nosepiece again to the high-dry (40×) objective. As you do so, watch the stage from the side to ensure that the slide is not touched as the objective clicks into place.

V Prepare a specimen drawing of a section of your specimen in Drawing Area B, including the internal structures. If the specimen in your field of view is one that you drew in Step R above, you may add the details to your outline.

Using the Oil-Immersion Power on Your Microscope

The oil-immersion (100×) objective allows you to view structures too small to see clearly with a high-dry objective. However, an oil-immersion objective is also more difficult to use, so you may need to practice using it several times before you become proficient. Follow these instructions exactly!

W Once you have focused the image on high-dry power, turn the nosepiece halfway toward the oil-immersion objective.

X Place one small drop of immersion oil on the center of the cover slip.

Y Turn the nosepiece so that the oil-immersion objective touches the drop of oil. Be sure to watch from the side to ensure that the objective doesn't hit the slide.

Z Using the fine adjustment knob, change the focus very slowly until you obtain the proper focus.

AA If the objective is raised so high that the oil separates from the objective, you have passed the point of focus. Repeat Steps Y and Z.

BB Observe a desmid or diatom using oil-immersion power and draw a portion of the specimen in Drawing Area C. Include the internal structures.

If you still cannot focus the image, you may have too much light. Adjust the light source or ask your teacher to adjust the diaphragm.

CC When you finish, clean the microscope and the slide carefully. Remove excess oil from the slide with a dry tissue. Clean the objective carefully with lens paper. (Again, never use anything but lens paper on a microscope lens.) Clean the slide with a wet tissue, being careful not to wet the label. Dry the slide thoroughly before returning it.

7. Scientists use scientific instruments to make better observations about the world. Write a paragraph on the ways that the microscope extends our ability to see things in our world.

_____

_____

_____

_____

_____

_____

_____

To ensure that your microscope lasts a long time, you should always store it properly. Carefully return the microscope to its storage place. Cover the microscope with a dust cover if one is available.

TABLE 1 Magnification of a Microscope

	Ocular Lens		Objective		Total Magnification
Scanning Objective	10×	×		=	
Low-Power Objective	10×	×		=	
High-Dry Objective	10×	×		=	
Oil-Immersion Objective*	10×	×		=	

*Some microscopes do not have an oil-immersion objective. If yours does not, skip this line of the table.

DRAWING AREA A

DRAWING AREA B

DRAWING AREA C

2 LAB

Lost in the Woods

Designing a Water Treatment System

You are alone and lost in the wilderness. You were with a group of people backpacking in the woods, but you fell behind and then took a wrong turn on one of the trails. A thunderstorm and a few blisters later, you discover that you are lost. You know that you need to stay in one spot to improve your chance of being found. You have plenty of food, a tent, and a light source. Temperatures are mild, so you don't have to worry about staying warm. Your biggest problem—drinking water. Thankfully, you've camped near a stream, but is the water safe to drink?

The water in natural areas often lacks the chemicals and pollutants that water in urban areas has. Trees and vegetation in watersheds act as natural filters for contaminants. But water from a stream can contain bits of gravel, sand, and soil. It can also contain bacteria and other critters that could make you sick.

The water you drink every day has likely been treated at a water treatment plant to remove contaminants that you would find if you scooped up water straight from its source, such as a local lake or river. People around the world need clean, fresh, disease-free drinking water. Let's explore how we can use changes in matter to purify water for drinking.

?

How can chemical and physical changes be used to purify water?

Equipment

laser pointer
universal indicator paper
petri dish and lid

nutrient agar
sterilized cotton ball
untreated water from a local water source

PROCEDURE

You are tasked with designing a water treatment system to filter a 250 mL sample of water. Your system must produce drinkable water as indicated by a series of water quality tests. You will be constrained by the types and quantities of materials that you can work with. As is always the case in real life, you will also be constrained by the amount of time that you have to design and build your system. Your teacher will specify a set of design parameters for you: the materials that you can use to construct your device and the time limitations for you to design and build it.

QUESTIONS

- What is dangerous about drinking unpurified water?

- How can you remove from water material that you can't even see?

- What tests can be done to verify the quality of water?

Planning the Design

A Start by researching how water treatment plants treat water from natural sources to make it drinkable. This may give you some ideas for your design.

B On the basis of what you have learned, design a way to treat water using supplies that could be carried in a backpack or found in the woods. Draw a diagram of your design on a separate sheet of paper. Indicate that this is a *preliminary design*.

C Share your research findings with your group. Reach consensus on a design that your group will build and submit for testing.

D Have someone in the group draw a full-size diagram of your group's agreed-upon design on a separate sheet of paper. You will use this during construction and submit a copy of it with your lab report. Indicate that it is the *initial production design*.

1. How will your mini water treatment facility eliminate particulates in water?

_____

_____

2. How will your device kill any bacteria in water?

_____

_____

Constructing the Treatment System

E Now build it! Pull together the materials you need and construct your water treatment device. Take a picture of your finished treatment system and paste it into Photo Area A.

Testing the Design

F Obtain a 250 mL water sample to be treated from your teacher. Time how long it takes the water to run through your device.

3. How long did it take?

_____

Now you need to prove that you have drinkable water—without drinking it! You will need to prove that there are no contaminants in your water.

G Test your water for particulates by shining a laser pointer through the water.

4. What did you observe about the clarity of your water?

_____

_____

_____

H Test your water for harmful chemicals by using indicator paper to test the pH.

5. Report the pH of the water that you tested.

_____

I Swab the agar in a petri dish with a sterilized cotton ball that has been dipped in your water sample to test your water for bacterial contaminants. Cover it with the lid and leave it undisturbed for two days.

6. Report the results of the bacterial test of your water.

_____

_____

_____

GOING FURTHER

7. Considering your treatment system's performance, what would you say are its strengths and weaknesses?

_____

_____

_____

_____

8. What changes could you make to your treatment system to improve its performance?

_____

_____

_____

J If you have the opportunity, redesign, rebuild, and retest your design. Submit a drawing of this as the *modified production design*.

9. How could you use the way that nature filters water in the environment in your own water treatment device?

_____

_____

10. The Greenville Water company owns 100% of the watershed that supplies Greenville, South Carolina, with water. In light of what you have learned about how trees and vegetation naturally filter water in a watershed, why do you think it is important to protect these areas from development?

_____

_____

_____

_____

_____

11. How is establishing protected areas an example of good and wise dominion through conservation?

_____

_____

_____

_____

PHOTO AREA A

3A LAB

Tag!

Mark-and-Recapture Sampling and Population Size

Fisheries biologists often need an idea of how many fish are in a pond or lake. They may need to determine, for example, whether a certain lake needs to be restocked with fish before fishing season opens. The most certain way to do this would be to catch and count all the fish in the lake, but this would be nearly impossible. So instead of fruitlessly trying to catch all the fish in the lake, biologists use a technique called *mark and recapture* to estimate the population size in a body of water. You will use the same technique to estimate the size of a population of crickets.

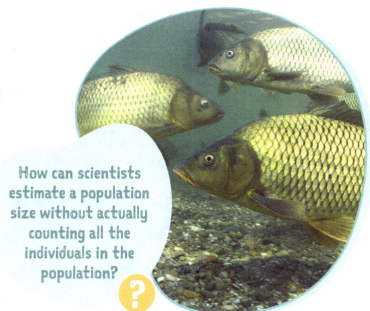

How can scientists estimate a population size without actually counting all the individuals in the population?

Equipment

terrarium or large bucket	dip net or cup
live crickets	correction pen

PROCEDURE

A From your terrarium or large bucket, randomly capture some crickets in a dip net or cup.

B Using the correction pen, mark each cricket that you've captured on the back of the thorax above the wings. See sketch on the right.

C Count the number of crickets as you mark them.

D Record the number of crickets that you marked in the first row of Table 1.

E Return the crickets to the terrarium and allow them to reintegrate into the population.

F After five minutes, randomly capture a second sample of crickets from the terrarium.

G Record the total number of crickets in your second sample in the second row of Table 1.

H Record the number of marked crickets in your second sample in the third row of Table 1.

QUESTIONS

- What is mark and recapture?

- How can mark and recapture be used to estimate population size?

- Is mark and recapture an accurate sampling method?

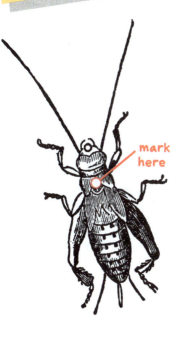

mark here

ANALYSIS

One way to analyze data from mark and recapture is to use a formula called the *Lincoln-Petersen method,* named for two biologists who studied large populations of fish and birds. The Lincoln-Petersen method begins with the proportion

$$\frac{n}{N} = \frac{k}{K},$$

where n = the number of organisms that were originally marked, N = total population size, k = the number of marked organisms that are recaptured, and K = the total number of organisms that are captured. In other words, this proportion says that the ratio of marked organisms to the total population (n/N) is the same as the ratio of marked recaptured organisms to the total number of captured organisms (k/K).

When solved for N, the proportion takes the form below.

$$N = \frac{Kn}{k}$$

1. Apply the Lincoln-Petersen method to your data in Table 1. What is your estimate of the total number of crickets? Show your work below.

_____

While the Lincoln-Petersen method tends to give accurate estimates for larger populations, it is less accurate for small populations such as the crickets in the terrarium. Douglas G. Chapman found that the method works much better if we add 1 to each factor.

First, the Lincoln-Petersen formula is set up as a proportion.

$$N = \frac{Kn}{k}$$

$$Nk = \left(\frac{Kn}{k}\right)k$$

$$N\cancel{k}\left(\frac{1}{\cancel{k}n}\right) = K\cancel{n}\left(\frac{1}{k\cancel{n}}\right)$$

$$\frac{N}{n} = \frac{K}{k}$$

$$\frac{n}{N} = \frac{k}{K}$$

Then, 1 is added to each factor.

$$\frac{n+1}{N+1} = \frac{k+1}{K+1}$$

Finally, the Chapman modification is solved for N.

$$N = \frac{(K+1)(n+1)}{(k+1)} - 1$$

2. Apply the Chapman modification to your data in Table 1. What is your estimate of the total number of crickets? Show your work below.

_____

3. For the Lincoln-Petersen method to be accurate, certain assumptions must be true. First, the population must be constant. What are some events that would make this assumption false?

_____
_____
_____

4. Is it likely that this assumption is true in this experiment? Explain.

_____
_____
_____
_____

5. A second assumption is that the animals captured and marked are just as likely to be captured the second time as the animals that have not been marked. Suggest a way of marking crickets that would violate this assumption.

_____
_____
_____

Since you have a captive population of crickets, you can calculate how close your estimates are to the actual population. A common way of determining the accuracy of an estimate is to calculate percent error using the formula below.

$$\% \ error = \left(\frac{|value_{estimated} - value_{actual}|}{value_{actual}} \right) 100\%$$

6. Calculate the percent error of your Lincoln-Petersen method estimate compared to the actual number of crickets provided by your teacher.

7. Calculate the percent error of your Chapman modification estimate compared to the actual number of crickets.

8. Which method was more accurate?

_____

GOING FURTHER

Now that you've finished using mark and recapture to estimate the number of crickets in your laboratory's terrarium, let's apply what you have learned to help your state fish and wildlife agency estimate the white-tailed deer population in a particular county.

The agency will use this fall's deer hunting season—when hunting of both bucks and does is allowed—as the recapture event by comparing the numbers of marked and unmarked deer killed. Propose a way to mark deer, keeping these requirements in mind.

» You must mark the deer in a way that will not make them more likely or less likely to be killed by hunters this fall.

» You must mark the deer in a way that will not spoil the meat. Hunters will not appreciate killing a deer only to find that they cannot eat the venison because the state agency adopted a poorly conceived sampling plan.

» You must mark the deer in such a way that the mark will not be lost or wear off in the time between now and the end of hunting season. (For simplicity's sake we will assume that the deer will not be shedding antlers or coats between now and then.)

» Finally, you must mark the deer in such a way that hunters can easily determine that the deer has been marked.

9. What is your plan?

_____

_____

_____

_____

_____

10. The agency put your plan into action and marked 2000 deer at the beginning of the hunting season. Out of 6578 deer killed during this year's deer season, 459 of them were marked. Using the Lincoln-Petersen method, estimate the deer population in this county at the end of the hunting season.

11. It would be difficult to ensure that no animals were added to or lost from the population during the length of this experiment. List two ways that individual deer could have joined or left the population.

_____

_____

_____

_____

12. Someone tells you, "Counting fish and deer is a waste of time. We should be spending our money on more important things, like people who are sick or homeless." How would you respond?

_____

_____

_____

TABLE 1

	ESTIMATE
n	
K	
k	

3B LAB

Must You Be So Competitive?

Inquiring into Growth Rate

If you were to walk through an older forest, you would probably notice that there were few young trees. Plants need sunlight to live and grow, and relatively little sunlight makes its way through the canopy formed by mature trees. Young trees tend to grow slowly in the forest's shady depths.

Sunlight is just one of the resources in an ecosystem, and the availability of resources governs how fast a population can grow. When conditions are ideal, the number of organisms in a population can soar. Conversely, if even one factor is limited, a population's growth may stagnate or even decline. In this lab activity you will examine how differences in an environmental factor affect plant growth. Duckweed is a small plant that usually reproduces asexually. This makes it an excellent test subject for population growth since you won't have to wait for it to flower or bother with pollinating it.

?

How does a plant's environment affect its growth rate?

Equipment
duckweed

QUESTION

- How do differences in environmental factors affect a population's growth rate?

PROCEDURE

Planning/Writing Scientific Questions

A Together with your team, decide on an environmental factor that you will test in relation to duckweed growth. Possible factors include but are not limited to the following.

» duration, intensity, or color of light exposure

» amount or type of fertilizer

» moving water versus still water

B Brainstorm with your lab group about how you can test the selected environmental factor while keeping other factors constant.

C Write specific questions related to the environmental factor that you could answer by collecting data.

Designing Scientific Investigations

D Write procedures to collect data that will allow you to answer the questions that you wrote in Step C.

E Have your teacher approve your procedures.

Conducting Scientific Investigations

F Collect data according to the procedures that you have written, then answer the questions that you wrote.

Developing Models

G Use your data to develop models of how changes in the environmental factor affect the growth of duckweed.

H Propose at least one way that you would change your experimental procedures to obtain better results.

Scientific Argumentation

I Explain what you know about the relationship between changes in the environmental factor that you have chosen and its effect on duckweed growth. Support your claims with evidence from the data that you collected.

4A LAB

Forest or Farm?

A Mathematical Model of Biodiversity

Imagine a rainforest. Towering trees draped in a tangle of vines, dotted with exotic orchids and bromeliads. Brilliantly colored macaws call raucously from the canopy while jewellike hummingbirds flit from flower to flower in search of nectar. Tapirs softly tread well-worn paths through the forest, keeping a wary eye out for the occasional jaguar. Caimans drift silently in the rivers, scattering schools of shimmering fish. And everywhere there is an astonishing variety of insects, from ants to beetles to butterflies.

Now imagine a cornfield—acres and acres of corn planted in neat rows. Thanks to herbicides and insecticides, the field is free of weeds and relatively free of insects. It's mainly corn—and not much else! That's the way the farmer wants it.

One of these two ecosystems has many different kinds of organisms living together in the same area. The other ecosystem consists almost entirely of only one kind of organism. Scientists say that the rainforest has a high biodiversity, while the biodiversity of the cornfield is low. Biodiversity is a measurement of the variety of life in a particular ecosystem. It's pretty easy to tell just by looking that the biodiversity of a cornfield is low, but what about other kinds of ecosystems? What about a forest, a meadow, or even a vacant lot near you? Assessing the biodiversity of an ecosystem gives biologists a valuable piece of information that can be used to help manage the system wisely. In this lab activity you will measure the biodiversity of an ecosystem by collecting and analyzing data with the help of a special formula from the biologist's mathematical toolbox.

? How do scientists decide whether an ecosystem is biologically diverse?

Equipment
field tape
field notebook
camera

QUESTIONS
- How do scientists measure biodiversity?
- How do scientists obtain representative samples from very large systems?

PROCEDURE

A Select a site for sampling, such as a forest, field, or other ecosystem.

1. Give a brief description of the site you've chosen to sample, including its location and the date of the sample.

_____

_____

2. Why do you think it is important, especially for field studies, to include the date of your sample?

_____

_____

3. Do you think the biodiversity of your sampling site is high, low, or somewhere in between? Write your answer in the form of a hypothesis.

_____

_____

B Lay your field tape in as straight a line as possible across your sampling area. This technique is known as *conducting a transect*—a method for collecting a subset of data from a much larger potential sample.

C Decide on the number of observations that you will make along your transect (each organism you tally is one observation). If your sampling area is small, you might choose to tally every organism that your transect touches. This is sometimes referred to as 100% sampling.

4. What is the advantage of a 100% sample?

_____

_____

_____

5. What is a possible disadvantage of collecting a 100% sample?

_____

_____

_____

_____

D If the sampling area is large, an alternative to a 100% sample is to conduct a smaller sample. For example, if your transect is 25 m long, you might decide to make an observation once every meter or half-meter along the length of the line.

6. What is the advantage of a smaller sample size (less than 100%)?

_____

_____

_____

7. What is a possible disadvantage of a smaller sample size?

_____

_____

_____

_____

E Starting at 0 m on your transect line, begin listing and tallying the kinds of organisms that your transect line touches. Count only those organisms that the line actually touches. Use Table 2 to list each new kind of organism that you encounter along your transect. Tally the number of each kind that you encounter. If you need additional rows, copy Table 2 onto a separate sheet of paper and attach it to your lab report.

ANALYSIS

The mathematical tool referred to in the introduction is a measure of biodiversity known as *Simpson's Index*. The formula for the index is

$$D = \frac{\Sigma n(n-1)}{N(N-1)},$$

where D is the index value, the Greek letter sigma (Σ) means "sum of," n is the number of individuals of any one species in the sample, and N is the total number of organisms in the sample. The result is a decimal number whose value is between 0 and 1.

F Calculate $n - 1$ and the product $n(n - 1)$ for each observed species and record in Table 2.

G Add up all the values for $n(n - 1)$ calculated in Step F. This is $\Sigma n(n - 1)$. Record this value in Table 2.

H Add up all the values for n; this sum is N. Record the value for N in Table 2.

I Multiply the total number of observed organisms (N) times that number minus one ($N - 1$). This is $N(N - 1)$. Record this value in Table 2.

J Calculate the value of Simpson's Index for your sample and record this value in Table 2.

K Simpson's Diversity Index, the actual number that you need to assess the diversity of your sample, is slightly different than Simpson's Index. To obtain the value of Simpson's Diversity Index for your sample, subtract the value that you calculated in Step J from 1 (i.e., $1 - D$) and record this value in Table 2.

CONCLUSIONS

So now you have a measure of the biodiversity of your sample, but what does the measure actually mean? To answer that question, let's think about the cornfield again. Imagine a cornfield with exactly one hundred corn plants and one dandelion in it and nothing else, the almost-perfect cornfield—hardly any weeds and no pests. That's definitely a lack of diversity! Plugging the values for the entire cornfield into the formula for Simpson's Index yields a Simpson's Index value of 0.98. See Table 1 and the calculation below.

TABLE 1 Cornfield Data

Species	Number (n)	n – 1	n(n – 1)
corn	100	99	9900
dandelion	1	0	0
	$N = 101$		$\Sigma n(n-1) = 9900$
	$N(N-1) = 10\ 100$		

$$D = \frac{9900}{10\ 100}$$

$$= 0.98$$

Subtracting 0.98 from 1 gives us a Simpson's Diversity Index value of 0.02.

8. What does the Simpson's Diversity Index value for the imaginary cornfield tell you about the relationship between biodiversity and the values produced by the index calculation?

   _____

   _____

9. What does the Simpson's Diversity Index value that you calculated for your sample tell you about the biodiversity of the ecosystem that you sampled?

   _____

   _____

10. Does the Simpson's Diversity Index value that you calculated for your sample support your hypothesis from Question 3? Explain.

    _____

    _____

    _____

    _____

11. Discuss some reasons why your Simpson's Diversity Index value may not be reflective of the actual biodiversity in the ecosystem that you sampled.

_____

_____

_____

_____

_____

12. How do mathematical tools such as Simpson's Diversity Index help biologists exercise wise dominion over God's creation?

_____

_____

_____

_____

_____

TABLE 2 Transect Data

SPECIES	NUMBER (*n*)	*n* – 1	*n*(*n* – 1)
	$N =$ _____		$\Sigma n(n - 1) =$ _____
	$N(N - 1) =$ _____		
	$D =$ _____		
	$1 - D =$ _____		

4B LAB

Hale Hardwoods or Sickly Cedars?

Monitoring Forest Health

Think of the last time you had the flu. You probably had some obvious and visible symptoms, such as a fever, runny nose, and chills. How can a biologist tell whether an ecosystem is sick and in need of some attention? The signs that an ecosystem is in trouble are not always obvious to an untrained observer. A forest, for example, that looks perfectly healthy to you may, in fact, be very "sick." But just as your family doctor can evaluate your fever and sniffles, scientists have ways to evaluate the health of ecosystems. They can determine whether an ecosystem is in prime condition and, if necessary, prescribe an appropriate plan of treatment to get their "patient" on the road to recovery.

When you go to the doctor's office, you don't get a prescription for medicine as soon as you set foot in the door. The doctor or an assistant first collects some basic information about you, such as your age, height, weight, temperature, pulse, and blood pressure. Biologists do the same thing when assessing an ecosystem—they first need some basic information. In this lab activity you'll try your hand at collecting an ecosystem's vital signs.

How can forest professionals give a forest a checkup?

Equipment

field tape

flagging tape

calculator

rangefinder

PROCEDURE

Establishing a Plot

Forest ecologists, dendrologists (scientists who study trees), and foresters (scientists who manage forest resources) often need to collect forest data. Forests, of course, have many trees in them, sometimes numbering into the millions. Assessing every tree would be impossible! What scientists need is a *sample* of the trees in the forest. Scientists call this a *plot*.

For this activity you will need a forest—or at least a place with some trees! A park or wood lot will do. The plot is a circular area representing a fraction of one acre. The size of that fraction is determined by the radius of the plot circle.

QUESTIONS

- How can scientists assess the health of a forest?

- What can an ecologist learn about a forest using what he knows about the size and number of trees in a sample?

- Why is it important to manage forest resources?

A Working with a partner, choose a plot large enough to include at least a few trees but not so large that you won't have enough time to assess all the trees within your plot. Use the chart below to help determine how large a plot to establish. Record your plot size in Table 1.

TABLE 1

PLOT SIZE (ACRES)	RADIUS (FT)
1/5	52.7
1/10	37.2
1/20	26.3
1/30	21.5
1/40	18.6
1/50	16.7

B Once you have determined how large a circle to make, use a field tape to mark the boundary of your plot. Have your partner hold one end of the field tape at the center of the circle while you measure out the appropriate radius. Mark the end of the radius with a piece of flagging tape.

C Repeat Step B enough times to sufficiently establish the boundary of your circle. When you are done, you should have an easily seen border between the trees inside the plot and those outside.

A 1/5 acre plot

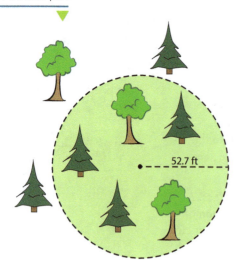

52.7 ft

Diameter at Breast Height (DBH)

It may sound silly, but we first have to define the word *tree*. Biologically speaking, every sapling in the forest is a tree, but young trees lack many of the features of mature trees. They don't have crowns, trunks, or very much woody tissue. Foresters define a tree as a woody plant with a trunk at least 3 in. in diameter at breast height, or DBH, where breast height is defined as 4.5 ft above ground.

D Use your field tape to measure 4.5 ft above ground on the trunk of one of the trees in your plot. Wrap the field tape around the trunk at that level. The point at which the 0 in. mark on the tape aligns with the wrapped tape indicates the circumference of the tree.

E Use a calculator to divide the measurement from Step D by 3.14. Since the circumference of the tree is equal to π times the diameter ($C = \pi d$), the result is the tree's DBH.

F If the DBH is 3 in. or greater, then that tree must be included in your sample. Record its DBH in Table 1.

G Repeat Steps D–F for each tree in your plot. Do not record data for any tree whose DBH is less than 3 in. Add more rows if needed.

Stem Count

One indication of forest health is *stem count*, or the number of trees (stems) per acre. To find the stem count represented by your sample, multiply the number of trees (as defined by DBH) in the sample by the number of plots it would take to make one acre. For example, if you have five trees in a 1/20 acre plot, then multiply 5 by 20 (5 trees per plot × 20 plots to make 1 acre), giving 100 stems per acre.

H Calculate the stem count per acre for your forest on the basis of your plot data and record this value in Table 1.

1. Why do you think stems per acre is an important indicator of forest health?

_____

_____

_____

_____

_____

2. Do you think that different species of trees have the same ideal number of stems per acre? Explain.

_____

_____

_____

_____

_____

Basal Area

Once you have determined that a tree in your plot is, in fact, a tree by a forester's definition, you next need to determine its basal area—a measurement of how much ground surface area the trunk of the tree covers. A shortcut for finding the basal area in square feet of a tree whose diameter is measured in inches is to take the square of the number of inches in its diameter and multiply the result by 0.005 454. For example, the calculation for a tree with a 6 in. diameter would be

$$(6 \text{ in.})^2(0.005\ 454 \text{ ft}^2/\text{in.}^2) = 0.2 \text{ ft}^2.$$

Thus, a tree with a diameter of 6 in. has a basal area of 0.2 ft^2.

I Calculate the basal area for each tree in your sample and record the results in Table 1.

3. What do you think the basal area for an individual tree can tell a dendrologist about the age and maturity of that tree?

_____

_____

_____

_____

Basal Area per Acre

Using the basal areas of the trees in your plot, you now need to determine how much basal area there is per acre in the part of the forest that you are assessing.

J To calculate your forest's basal area per acre, start by multiplying the basal area for each tree in your plot by the value of the denominator of your plot size. For example, the basal areas for trees in a 1/20 acre plot would each be multiplied by 20. Do this for each tree in your plot and record the results in Table 1.

K Add up all the results from Step J to get the calculated value for the total basal area per acre for your forest and record this total in Table 1.

4. What sorts of information about forest health can a forest ecologist learn from basal area per acre? (_Hint:_ Think about what might be indicated by a small basal area versus a large one.)

_____

_____

_____

_____

_____

_____

_____

_____

_____

_____

5. Why would a forester need to know _both_ the stem count per acre and basal area per acre for a particular forest?

_____

_____

_____

_____

_____

CONCLUSIONS

6. On the basis of the stem count and basal area per acre that you calculated for your forest, what conclusions can you make about the age and maturity of the trees in your forest?

_____

_____

_____

_____

7. What aspects of this lab activity might indicate that your conclusions in Question 6 are unlikely to be very reliable? What could you do to improve the reliability of your conclusions?

_____

_____

_____

_____

8. What other kinds of data could a forest ecologist collect? Describe what that data might say about the health of a forest.

_____

_____

_____

_____

_____

GOING FURTHER

Ponderosa pine is a large conifer (cone-bearing tree) native to the American West. Ponderosa pine seeds grow best on bare soil and in full sunlight. The life cycle of these trees is very closely bound to the periodic wildfires that regularly occur in their habitats. Fires clear the ground of competing trees and shrubs, return nutrients to the soil, and cause ponderosa pine seed cones to open and release their seeds.

9. Would stands of ponderosa pine tend to be more evenly aged (being trees of similar age) or unevenly aged (being trees of widely varying age)? Explain.

_____

_____

_____

10. A particular stand of ponderosa pines is found to have a low stem count and high basal density. What can you deduce about the amount of shade in this stand? Would this be a good place for ponderosa seedlings to grow?

_____

_____

11. Other than timber products, such as lumber, what other vital goods and services do forests provide?

_____

_____

_____

12. How does understanding forest health fit into a Christian worldview?

_____

_____

_____

_____

_____

_____

_____

_____

A stand of ponderosa pines

TABLE 1

PLOT SIZE (ACRES):			
Tree	DBH (in.)	Tree Basal Area (ft²)	Basal Area per Acre (ft²/acre)
1			
2			
3			
4			
5			
6			
7			
8			
9			
10			
11			
12			
Stem Count/Acre		Total Basal Area/Acre	

5A LAB

Dwell on the Cell

Basic Cytology

Imagine that you are a biology student in the year 1600. The microscope has not yet been invented, and no one, including the greatest scientific minds of the day, has a clue about the existence of an entirely unseen microscopic world all around them—and on them, even in them! Like the children in *The Chronicles of Narnia* who stepped through the wardrobe, the first microscopists to peer through their primitive, homemade lenses would encounter a world that defied imagination.

In Lab 1A you learned how to use a microscope. In this lab activity you will learn an additional microscopy technique: the making of a wet mount—a temporary microscope slide in which the specimen is mounted in water or some other fluid. After you have read about how to prepare a wet mount, you will practice the technique by doing what Robert Hooke did to observe a slice of cork. You will then observe some typical plant and animal cells, in the form of onion epidermal cells and human cheek cells.

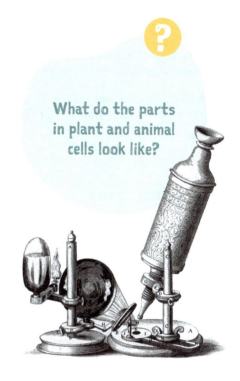

What do the parts in plant and animal cells look like?

Equipment

microscope	pipette
dissection kit	onion
cork stopper	methylene blue
hexagonal metal nut, large	paper towels
razor blade, single-edged	flat toothpick
glass slides (3)	laboratory apron
cover slips (3)	goggles

QUESTIONS

- What did Robert Hooke see when he looked at cork through a microscope?

- What does a typical plant cell look like?

- What does a typical animal cell look like?

PROCEDURE

Observing Cork Cells

Over three hundred years ago Robert Hooke discovered that certain plant tissues are made up of what he called *cells*. To get a proper perspective of cytology, you will repeat his experiment. To see cork cells well, you must use a very thin slice of cork, only one to two cells thick. If suitable slices of cork are provided for you, skip Steps A and B.

A Take a small cork stopper (piece of cork) and insert it into a hexagonal metal nut. Twist it carefully so that the flat surface of the cork does not become crooked inside the nut. As the cork reaches the other side, continue to turn it until it barely protrudes beyond the nut.

B Run a single-edged razor blade along the surface of the nut, carefully cutting into the cork. You do not need to get an entire cross section of the cork, but the section you use must be very thin.

C Prepare a wet mount of cork by following the instructions in Appendix D. *Note:* If your cover slip teeters on the cork, your slice of cork is too thick. Try making a thinner slice.

D Observe the wet mount of cork on low power. Your slice is probably thinnest along one of the edges, so you might want to start exploring there first.

1. When placing wet mounts on the microscope stage, the slide must remain parallel to the floor. Why is this necessary?

_____

_____

_____

2. Can you see any internal cellular structures in the cork cells? Explain.

_____

_____

3. You are, of course, observing dead cork. What cellular structure are you observing?

_____

Observing Onion Epidermal Cells

E Obtain the scale of a small onion. (A scale is one of the onion's layers.) The epidermis of the onion is a thin, translucent skin on the inside surface of the scale. Take the scale and break it. At the edges of the broken scale you should be able to see a portion of the epidermis.

F Peel a translucent layer from the scale using forceps. (A translucent layer allows light through but is not transparent.) The epidermis is a very thin sheet of cells, so do not crush or wrinkle it. If you do, the cells can be damaged and air bubbles will get trapped between the layers, making them hard to observe.

G Prepare a wet mount using a small piece of onion epidermis no larger than the drop of water on your slide. Place the onion epidermis so that it lies flat. If it begins to fold or curl, use probes to straighten it. Put a second drop of water on it and then put the cover slip on. The cover slip should adhere tightly. If it appears to be floating, you can draw some water off with a paper towel. If there are large air bubbles, then there is not enough water, and you can add small amounts with a pipette at the edge of the cover slip. You may need your lab partner's help.

H Observe the onion epidermis cells under low power.

4. What is the general shape of one onion epidermal cell?

_____

_____

5. What is the general shape of a group of onion epidermal cells?

_____

_____

6. What do the cork and onion cells have in common?

_____

_____

_____

7. Of the following terms, indicate the ones that apply to the onion epidermis: unicellular, multicellular, tissue, organ system.

_____

_____

I Carefully remove the slide from the stage and place it next to the microscope. Stain your onion epidermal cells by placing one drop of methylene blue on the slide at the very edge of the cover slip, in contact with the water under the cover slip. At the opposite side, touch a paper towel to the water under the edge of the cover slip, allowing the paper to absorb the water.

The stain will be drawn under the cover slip. (If the stain runs over the outside edges of the cover slip, you probably used too much water when you made your wet mount. Use paper towels to absorb the excess water and try again.) When the stain has contacted the onion epidermis, blot away any excess fluids on the slide or cover slip.

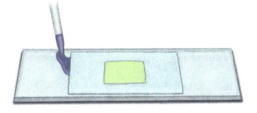

J Allow the stain to remain on the slide three to five minutes before observing the specimen. This permits the stain to enter the cells.

K Observe the stained onion epidermis on low and on high power. If you don't see any changes, you may need to add another drop of stain or wait a little longer for the stain to have an effect.

8. What can you see that differs from your observation of an unstained onion epidermis?

_____

_____

_____

9. Look among the cells until you find a dark, circular structure inside one of them. What is it?

_____

10. Frequently, darker spots can be seen within this dark structure. What are they?

_____

L In Drawing Area A make a drawing of one onion cell with a few adjoining cells to show how the cells fit together. Use the power that you feel is best, but do not use the oil immersion lens. Draw the internal structures for the main cell only. Label only the structures that you see in your specimen. Be sure to indicate which power you used to prepare your drawing.

Observing Human Cheek Epithelial Cells

M Prepare a wet mount of your cheek cells or those of your lab partner. Collect some mucous epithelial cells by rubbing the blunt end of a flat toothpick back and forth inside your cheek. ***Collect cells from your own mouth only!*** To get the greatest concentration of cells, do not twirl the toothpick around; use only one side. Remove the toothpick carefully, collecting as little saliva as possible.

N Put one drop of methylene blue on the center of a microscope slide. Immediately tap the edge of the toothpick with the cells several times in the stain. After this is done, carefully add the cover slip.

11. Why do you need to be careful when you place the cover slip on top?

_____

_____

_____

_____

O View the cheek cells under low power. Look for isolated cells, not clumps.

12. How can you distinguish the epithelial cells from the other debris that appears on the slide?

_____

_____

_____

13. These cells are called *mucous epithelial cells. Mucus* (noun form of *mucous*) is the slimy substance produced by membranes to moisten and protect them. What does the word *mucous* tell you about the functions of these cells?

_____

_____

_____

P Look for a suitable cheek epithelial cell to draw, center it in your microscope's field of view, focus on high power, and then draw the cell in Drawing Area B. Label all the parts that you see.

14. What are some of the similarities between the onion epidermal cells and the cheek epithelial cells?

_____

_____

15. What are some of their differences?

_____

_____

CONCLUSION

16. Why don't cheek cells all have the same shape on the slide?

_____

_____

_____

17. If you were to stain both types of cells, which would stain more quickly? Why?

_____

_____

_____

18. What is the purpose of the plant cell wall?

_____

_____

GOING FURTHER

19. Animal cells do not have cell walls since animals have other means of physical support, such as a skeleton. Discuss the problems that animals would have if their cells did have cell walls.

_____

_____

_____

_____

20. How does God's design of the cell for both plants and animals show His care for creation?

_____

_____

_____

21. David thanked God in Psalm 139:14 for being fearfully and wonderfully made. On the basis of your own understanding of cell structure, write a sentence or two expressing thanks to God.

_____

_____

_____

_____

DRAWING AREA A

DRAWING AREA B

5B LAB

The Pressure Is On

Investigating Osmosis

As we have learned, osmosis is a form of passive transport that helps cells regulate their internal environment. In this activity you will investigate how a difference in one factor—solute concentration—can affect osmosis. Afterward, you will extend what you have observed to other factors that might affect osmosis.

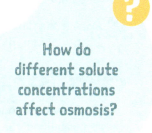

How do different solute concentrations affect osmosis?

Equipment

osmometer

squeeze bottle

sucrose solutions

wax pencil

ruler

paper towels

distilled water

laboratory apron

goggles

QUESTIONS

- Is the rate of osmosis affected by solute concentration?

- Does osmosis ever reach an equilibrium?

- Can I predict an effect on osmosis on the basis of solute particle size?

PROCEDURE

Solution A has twice the concentration of sucrose as Solution B. Your teacher will task you with testing one of the two solutions.

A Gently shake the osmometer bulb to remove all water.

B Use a squeeze bottle to completely fill the bulb with the assigned solution, then set the bulb on a wet paper towel.

C Uncap the squeeze bottle and insert the bottom 5 cm of the osmometer's tube into the solution. Place your thumb over the top end of the tube to keep the solution in the tube as you transfer it to the osmometer bulb.

D While you keep your thumb over the osmometer tube, place the tube into the neck of the osmometer bulb. Tap the bottom of the membrane to remove any air trapped in the bulb. Make sure that there is no air trapped in either the bulb or the tube.

E Use a paper towel to wipe the joint between the bulb and tube. If it leaks, ask your teacher for help.

F Rinse the outside of the bulb with distilled water.

G Fill the osmometer's beaker with distilled water to a depth of 5 cm.

H Suspend the osmometer bulb in the beaker so that the membrane is covered but the water does not come near the joint of the bulb and tube (your teacher will demonstrate the method to use). Use the wax pencil to mark the level of solution in the tube.

I After ten minutes have passed, measure how far the fluid has risen in the tube. Remember to measure from the fill mark, not from the bottom of the tube or bulb. Record your measurement in Table 1.

J Repeat Step I four times.

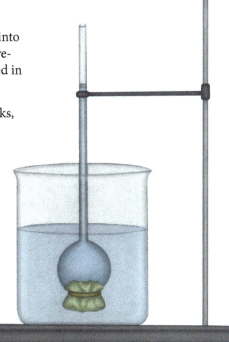

ANALYSIS

K Pool your data together with the results from the other lab groups in your class. Record the pooled data in Table 2. Remember to tabulate the results for Solution A and Solution B separately!

L Average the class data for Solution A and enter the result in Table 2. Repeat for the Solution B data.

CONCLUSION

1. On the basis of your pooled class results, what can you conclude about the permeability of the membrane to sucrose?

_____

_____

2. Is there a difference in the height gain between Solutions A and B? If so, what could account for the difference?

_____

_____

_____

GOING FURTHER

3. If you were to let the experiment run long enough, could all the water in the beaker enter the osmometer? Defend your answer using data from the procedure.

_____

_____

_____

_____

_____

4. If you were to slowly heat the water in the beaker, what might happen to the rate of osmosis? Explain.

_____

_____

_____

5. If you were to set up the osmometer using salt instead of sucrose, would you expect similar results? Why or why not?

_____

_____

_____

_____

_____

6. Sucrose is a *disaccharide*, a sugar made of two different, smaller molecules, glucose and fructose. Invertase is an enzyme that decomposes sucrose into glucose and fructose. Predict how your test results might be different if invertase were added to your sucrose solutions.

_____

_____

_____

_____

_____

TABLE 1 Test Results

Solution (A or B)	COLUMN HEIGHT (mm)					
	Start	10 min	20 min	30 min	40 min	50 min
	0 mm					

TABLE 2 Class Results

Solution (A or B)	COLUMN HEIGHT (mm)					
	Start	10 min	20 min	30 min	40 min	50 min
	0 mm					
	0 mm					
	0 mm					
	0 mm					
	0 mm					
	0 mm					

Solution	AVERAGED COLUMN HEIGHT (mm)					
	Start	10 min	20 min	30 min	40 min	50 min
A	0 mm					
B	0 mm					

6A LAB

No Swimming Today

Oxygen and Metabolism

It's a hot, sunny summer day, and you are looking forward to jumping into the pool and feeling the cold water close over your head. You get to the pool and notice that there is no one around. A sign says, "Pool closed today—dangerous bacteria levels."

Public pools use tests to make sure that people are not swimming in water that could make them sick. Pool water must be treated to ensure that algae and bacteria don't grow in it. Bacteria can feed on organic matter in a swimming pool, and even if chlorine kills the bacteria, the water could still be polluted.

Scientists and wastewater technicians test the water we drink to make sure that it doesn't contain organic materials. When a large amount of organic material enters a body of water, the bacteria that normally decompose dead plants and other organic materials increase. This is a problem because the bacteria use oxygen to break down these organic materials, causing the oxygen levels in the water to plummet. The problem is especially bad at night when plants are not producing oxygen through photosynthesis.

When working properly, wastewater treatment facilities remove most of the organic matter before the treated water is released into a body of water. However, technicians constantly monitor the body of water to make certain that treated water is clean. One of the ways they do this is to use a test to measure *biochemical oxygen demand* (BOD)—the oxygen needed for bacteria to break down organic matter in a water sample at 20 °C over a period of time. It's a way to measure how much organic matter is in a water sample. A crystal-clear stream has a BOD of 1 mg/L of water over a five-day period. Untreated sewage in the United States has an average BOD of 200 mg/L for the same time period.

You will be measuring the BOD of a water source by comparing the dissolved oxygen content of the water right after it was collected with the oxygen content of the water five days after it was collected. This waiting period allows bacteria in the water to use the oxygen to break down any organic matter.

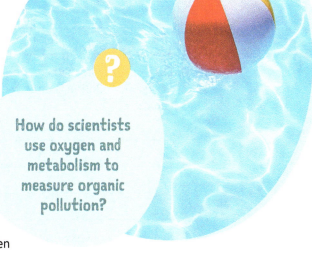

How do scientists use oxygen and metabolism to measure organic pollution?

QUESTIONS

- What is biochemical oxygen demand (BOD)?

- What is the link between oxygen and metabolism?

- How do we test for BOD?

- How do we use BOD to estimate the amount of organic pollution in a water sample?

Equipment

LaMotte 5860-01 test kit
two water samples
nitrile gloves

laboratory apron
goggles

1. Why do bacteria need oxygen to metabolize waste products?

_____

_____

2. Why does the temperature for BOD need to be specified?

_____

_____

3. Why does the length of time for BOD need to be specified?

_____

_____

_____

4. Do you think the mathematical relationship between BOD and the amount of organic pollution is directly or indirectly proportional? Explain your choice.

_____

_____

_____

_____

5. Suggest how the oxygen that an organism requires for metabolism is related to its size, such as the difference in the metabolism of an elephant versus that of bacteria. Explain your reasoning.

_____

_____

PROCEDURE

Fixing the Sample

People who check water sources for safe bacterial levels usually use a BOD kit similar to the one shown on page 49. The first step will be to fix the water sample. Fixing means to treat the sample with chemicals that will lock in the oxygen that it already contains so that it doesn't change over time.

A Obtain a water sample by allowing water to flow into a dissolved oxygen (DO) bottle. Recap the bottle while it is submerged, then take the bottle out of the water. Label this sample as *Day 1*. Repeat this step to obtain a second sample and label it as *Day 5*. Store the Day 5 sample in a darkened area at room temperature until ready to test.

6. Where did your water samples come from? When did you obtain it?

_____

_____

7. Describe your observations of the Day 1 water sample, including its cloudiness, or *turbidity*.

_____

_____

8. On the basis of your observations, predict whether your water sample will have a BOD more similar to a clear stream (1 ppm) or sewage (200 ppm).

_____

9. Why is it important that your DO bottle not contain any air bubbles?

_____

_____

B Uncap the bottle. Put 8 drops of manganous sulfate solution and 8 drops of alkaline potassium iodide azide into the water sample. Since the bottle is already full, some water will probably spill over the side of the bottle.

C Recap the bottle and mix its contents by gently inverting the bottle several times. Let the water sit for several minutes.

10. Describe your observations of the water in the DO bottle.

_____

_____

D Uncap the bottle and add 8 drops of sulfuric acid. Recap and gently invert the bottle several times until the cloudiness, or precipitate, begins to disappear.

11. What did you observe about the water sample in the bottle?

_____

_____

Notice that the second chemical you added to the water sample was alkaline potassium iodide azide. As you have added chemicals, free iodine has been produced in solution. This is similar to a dark brown solution known as *tincture of iodine* that is sometimes used to treat wounds. The word *alkaline* means that this compound acts as a base.

12. Why is it helpful to react an acid with this basic solution?

_____

_____

Measuring Dissolved Oxygen

Now that the water sample is fixed, the dissolved oxygen in it is stable and can be measured. You will use a process called *titration* to find out how much dissolved oxygen is in your sample. Titration is a way to chemically react something of an unknown concentration with something of a known concentration. We can measure the volume of the solution of known concentration we use to determine the concentration of the unknown substance. You will use a titration syringe to measure this volume.

You are trying to find the concentration of oxygen in your solution. The sodium thiosulfate ($Na_2S_2O_3$) in the test kit is a solution whose concentration is known. This chemical will react with the iodine in the water sample according to the reaction below.

$$2Na_2S_2O_3 \ (aq) + I_2 \ (aq) \longrightarrow Na_2S_4O_6 \ (aq) + 2NaI \ (aq)$$

13. What does the (*aq*) in this reaction mean?

_____

_____

Scientists usually use an indicator that makes the solution change color so that we can know when the reaction is done. You will use a starch solution to determine that the reaction above is done.

Note: The titration syringe provided in some BOD kits is marked in parts per million (ppm). Parts per million is a ratio, not a unit of volume; in this instance, each 0.1 mL of sodium thiosulfate used during titration corresponds to 1 ppm of dissolved oxygen in the water sample. Instead of measuring the amount of titration solution used and then using that information to calculate the amount of dissolved oxygen in the sample, the dissolved oxygen level is read directly from the titration syringe. This method may vary; follow the directions provided with the kit you are using.

E Pour some of your fixed water sample into the titration bottle to the 20 mL mark.

F Using your titration syringe, draw up sodium thiosulfate to the 0 ppm mark. (The syringe is marked in reverse because you will be concerned with the amount dispensed, not the amount drawn up.)

G Deliver sodium thiosulfate into the titration tube drop by drop by putting the syringe in the hole in the lid of the titration tube. You may not need to use all the sodium thiosulfate solution, so it's important to add it one drop at a time. Swirl the titration tube between drops. Keep adding sodium thiosulfate until the solution color is about the same shade of yellow as a dandelion. Placing a piece of white paper behind the titration tube will help you see the color change better.

H Now carefully remove the titration syringe and gently set it upright in its storage tube. Uncap the titration tube and place the cap on your bench top. Add 8 drops of starch solution to the titration tube.

14. What did you observe when you added the starch solution to the fixed water sample?

_____

_____

Starch indicator changes color in the presence of free iodine. When all the free iodine is used up, the solution will quickly change to a clear color.

I Put the cap and syringe back on your titration tube and add sodium thiosulfate drop by drop, swirling the tube gently after each drop, until the solution flashes clear. If you use up all the sodium thiosulfate in the syringe, add more, keeping track of how much you've used. Once the solution flashes clear, record the corresponding amount of dissolved oxygen in parts per million (ppm) for Day 1 of Table 1. (See note prior to Step E.)

15. You will now let the water sample sit for five days and repeat the DO test. Predict whether dissolved oxygen in the second water sample will increase or decrease in storage. Explain your reasoning.

_____

_____

_____

_____

J On Day 5 follow Steps E through I to fix and measure the dissolved oxygen for the Day 5 sample. Record the amount of dissolved oxygen for Day 5 in Table 1.

ANALYSIS

16. Comment on the accuracy of your prediction in Question 15.

_____

_____

_____

K Subtract the DO value for Day 5 from the value for Day 1. This is your BOD value. Record the BOD in Table 1.

17. What is the BOD for your water sample?

_____

18. Using the information below, describe the nature of your water samples.

  » 1–2 ppm—very clean, little organic matter

  » 3–5 ppm—moderately clean, some organic matter

  » 6–9 ppm—poor water quality, abundant organic matter

  » ≥10 ppm—very unhealthy water, large amounts of organic matter

_____

_____

19. On the basis of the information about the water source that you recorded in Question 6, explain why your water had this cleanliness rating.

_____

_____

_____

20. How can we protect the environment to keep water sources around us as clean as possible?

_____

_____

21. Explain from a biblical worldview why we should keep water clean.

_____

_____

_____

GOING FURTHER

22. Design an experiment to test different factors that affect BOD.

_____

_____

_____

_____

_____

_____

_____

_____

_____

23. How could this information be useful to a county that is considering building a public park?

_____

_____

_____

_____

_____

_____

_____

_____

_____

TABLE 1

COLLECTION DAY	DISSOLVED OXYGEN (ppm)	BOD (ppm)
Day 1		
Day 5		

6B LAB

Hidden Code

Extracting DNA from Cells

A long-lost uncle has left you $1,000,000 in his will. All you need to do is to prove that you are his relative. But how are you going to do that?

You've been learning about how every cell stores information in the form of DNA. Scientists can extract DNA from any cell, including the cells in your body. In Chapter 9 you'll learn more about DNA fingerprinting, or DNA profiling. This is a process of forensics that involves extracting DNA from two individuals, cutting the DNA up into fragments, and comparing them to show that two people are genetically related. Because of the code hidden in your cells, you would be able to prove that you are an heir!

In this lab activity you will be extracting DNA from plant cells to see what DNA looks like and what scientists can learn from it. It's easier to extract and see DNA from the cells of some plants because their cells contain multiple copies of their DNA. Botanists can use the DNA in plants to figure out how to breed plants that have certain properties that make them more useful for people.

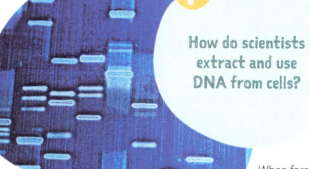

How do scientists extract and use DNA from cells?

When forensic scientists do DNA fingerprinting, the first step is to extract DNA from a person's cells. They get these cells by swabbing the inside of the person's cheek. The DNA is then cut up into fragments that can be observed and compared.

Equipment

blender
microscope
beakers, 150 mL (2)
graduated cylinder, 100 mL
stirring rod
microscope slide and cover slip
beaker, 50 mL

lentils, dried and uncooked
liquid dish detergent
isopropyl alcohol (rubbing alcohol)
methylene blue solution
salt solution, 6%
phenol red indicator solution

nitrile gloves
laboratory apron
goggles

QUESTIONS

- What substances contain DNA?
- How do scientists get DNA out of a substance?
- What does DNA look like?
- How do scientists use DNA to learn about living things?

1. Where will the DNA that you are extracting come from? Where will it not come from?

_____

_____

_____

2. Why do the cells in plants have DNA?

_____

_____

_____

3. Why must the plants that you are using be uncooked? (*Hint*: Think about the effect of heat on large molecules.)

_____

_____

_____

_____

PROCEDURE

A Add 60 mL of water to one of the 150 mL beakers.

B Using the graduated cylinder, measure 15 mL of lentils. Pour the lentils into a second 150 mL beaker. Add 45 mL of water to this beaker. Pour the mixture into a blender and blend on high for about 15 seconds until the lentils form a kind of sludge. Deposit this back into the beaker.

4. Why do you need to add water to the lentils?

_____

_____

_____

5. Why do you need to blend the lentils?

_____

_____

_____

_____

6. What do you observe about the contents of both beakers?

_____

_____

_____

_____

7. You will be adding chemicals to both beakers. Think about the way that an experiment is conducted. What is the function of the first beaker?

_____

_____

_____

_____

C Now add 8 mL of liquid dish detergent to both beakers. Gently swirl both beakers in a way that minimizes the amount of foam that forms on top.

8. Suggest why we are treating the lentils with soap, which will attract proteins in the cell membrane.

_____

_____

_____

_____

9. After adding the soap, where is the DNA?

_____

D While tilting the first beaker, add 45 mL of isopropyl alcohol to it, pouring the alcohol gently along the side of the beaker so that the alcohol forms a separate layer from the water. Do the same for the second beaker. Let both beakers sit for two to three minutes.

10. What do you observe about the contents of both beakers?

_____

_____

_____

_____

_____

_____

If you see bubbles rising up through the alcohol in the second beaker, observe them carefully. They may be drawing white strings of thousands of DNA molecules along with them as they rise. You will physically remove one of these DNA samples from the beaker.

E Insert a stirring rod all the way to the bottom of the second beaker. Slowly turn the rod one direction, winding any white strands that you observe around the rod. Gently remove the rod to observe the DNA strands. Clean the stirring rod, then repeat this procedure with the first beaker to see whether you can observe any DNA.

11. Why did two layers form in the beakers?

_____

_____

_____

F Take some DNA out of solution and put it on a microscope slide to prepare a wet mount. Add 2 drops of methylene blue and wait a few minutes for it to dye the DNA. Cover it with a cover slip.

G Using the microscope, view the DNA first at low power, then at a higher power.

12. What do you observe about DNA under a microscope?

_____

_____

_____

H Now measure 15 mL of 6% salt solution into a 50 mL beaker. Wind more DNA around your stirring rod and place it in this solution, turning the rod to dislodge the DNA.

13. What do you observe when you place the DNA molecules in the salt solution?

_____

_____

_____

Phenol red is an indicator that works on solutions that have a pH between 6.8 and 8.2. It will turn yellow in the presence of an acid, orange at a neutral pH, and pink in the presence of higher pH.

6.0   7.0   8.0

I Add 5 drops of phenol red indicator to the salt solution containing the DNA that you have dislodged.

CONCLUSION

Let's consider all the information that you have gathered to learn a little more about what DNA is like as a chemical.

14. Is DNA an acid, base, or neutral? How can you tell?

_____

_____

_____

15. What does DNA physically look like?

_____

_____

_____

16. The first beaker should contain no cells and therefore no DNA. If you were able to extract DNA from this beaker, what would you conclude?

_____

_____

_____

_____

17. The DNA that you observed is clumped together, but it will readily dissolve when it is stirred in water. Is DNA a mostly polar or nonpolar molecule? Explain.

_____

_____

_____

_____

18. You couldn't see a single strand of DNA, even though DNA is a large molecule over 2 m long. Why can you see DNA in your lentil solution?

_____

_____

_____

_____

19. When geneticists do DNA fingerprinting or analyze the genome of a plant that they are trying to breed, the first step is to extract the DNA. The second step is to cut the DNA up into fragments. Considering what you know about DNA, why do you think they do this?

_____

_____

_____

_____

_____

GOING FURTHER

20. How could a geneticist move a section of DNA that coded for something useful from one organism into another organism of the same species?

_____

_____

_____

_____

_____

One DNA molecule is over 2 m long!

21. The white goop that you extracted from the lentils includes both RNA and DNA because the process you used extracts all nucleic acids. Design an experiment that you could use to answer this question: What percentage of the lentils' mass is the mass of the nucleic acids?

_____

_____

_____

_____

_____

_____

_____

_____

7A LAB

Whatever Floats Your Leaf

Rates of Photosynthesis

Ah, Thanksgiving dinner! A turkey baked to golden perfection, creamy mashed potatoes, cranberry sauce—and Grandma's famous green bean casserole! You probably don't think too much about it, but photosynthesis makes all these delicious entrées possible. Without photosynthesis, not only would there be no Thanksgiving dinner—there'd be no dinners at all! And that would make for a sad sort of world, not to mention one in which God's creatures could not live.

Photosynthesis is a series of enzyme-catalyzed reactions that take place in autotrophic organisms such as green plants. Plants use this process to convert energy from the sun to stored energy in sugars that they can then use for cell growth. The leaves of a plant are the main photosynthetic factories. They contain chloroplasts with chlorophyll, which absorbs light energy. The overall formula shows the raw materials that plants need to produce their food.

$$6H_2O + 6CO_2 + \text{light energy} \longrightarrow C_6H_{12}O_6 + 6O_2$$

This one process provides all the energy needed for growth and survival for producers, like potato and green bean plants, and for consumers, like turkeys. Even secondary consumers, like people, ultimately get their energy from photosynthesis. In this lab activity you will use parts of leaves to demonstrate the photosynthetic process.

What factors can affect the rate of photosynthesis in plant leaves?

Equipment

hole punch	fresh spinach or ivy leaves	light source
oral syringe, 10 mL or larger	sodium bicarbonate solution, 2.5%	stopwatch
forceps	clear plastic cups	colored pencils

PROCEDURE

A Technique to Measure Photosynthesis

PREPARING THE LEAF DISKS

Parts of leaves normally float in solution since they are filled with oxygen and carbon dioxide, but they will sink when infiltrated with a sodium bicarbonate solution. As seen in the photosynthesis equation, leaves that undergo photosynthesis produce oxygen that is released into the leaf spaces, making a leaf part capable of floating. Respiration, which consumes oxygen, is also taking place in the leaves. The measurement of leaf parts rising is an indirect way of quantifying the net rate of photosynthesis.

A Use the hole punch to cut about fifteen disks out of a spinach or ivy leaf. Avoid punching out the leaf veins.

QUESTIONS
- How can we measure the rate of photosynthesis?
- What factors might affect the rate of photosynthesis?

B Remove the plunger from the syringe and gently load the disks into the bottom of the syringe. You may need to use forceps to gently push the disks to the bottom of the syringe. Replace the plunger in the syringe and push it down as far as it will go without squashing the disks.

C Pull 10 mL of 2.5% sodium bicarbonate solution into the syringe. Hold the syringe vertical with the tip up. Tap the syringe to get as many air bubbles as possible to rise to the tip and be released. Gently depress the plunger to force any remaining air from the syringe.

D Cover the syringe opening and pull back on the plunger to create a vacuum in the syringe. Maintain the vacuum for at least 10 seconds while shaking the disks to suspend them in the solution. Tiny air bubbles should be seen at the edge of the disks where the air is being pulled out of the leaf disk spaces. Tap the syringe to release the air bubbles from the disks. The leaf disks should begin to sink.

E Now depress the plunger while covering the syringe opening. This will force the bicarbonate solution into the leaves.

You may have to repeat Steps D and E several times to get all the leaf disks to sink.

1. Why is a solution of sodium bicarbonate ($NaHCO_3$) forced into the leaves instead of just tap water or distilled water? (*Hint*: Review the chemical equation for photosynthesis and think about what sodium bicarbonate might supply for the leaf disks.)

_____

_____

_____

Once the leaf disks have sunk, they are ready to be used for rate of photosynthesis trials.

MEASURING THE RATE OF PHOTOSYNTHESIS

F Pour the disks from the syringe into a clear plastic cup.

G Add 2.5% sodium bicarbonate solution to the cup to a depth of 2 cm. Separate the leaf disks so that they are not overlapping. Discard any disks that do not sink.

H Place the cup under a light source.

I Observe the number of leaf disks at the surface after each minute for 25 minutes. Record your observations in Table 1.

2. Describe what you observe.

_____

_____

_____

Testing Conditions for Photosynthesis

In this part of the activity you will design an experiment to use the leaf disk technique to test how varying factors can affect the rate of photosynthesis.

J Choose an environmental factor to test. Some suggestions are light intensity or wavelength, pH, and temperature. Note that these are not the only factors you might test. Check with your teacher for approval if you think of something not on this list.

3. What environmental factor have you chosen to test? Write your answer in the form of a research question.

_____

_____

_____

4. What is your hypothesis?

_____

_____

_____

5. Explain why you think the factor you've chosen to test affects the rate of photosynthesis.

_____

_____

_____

6. What is the independent variable in your experiment?

_____

7. What is the dependent variable in your experiment?

_____

8. What will be the standardized variables in your experiment?

_____

9. What will you use as a control group?

_____

10. Describe the procedure that you will use to test your hypothesis.

_____

_____

_____

_____

_____

K Run your experiment.

L Record your data in Table 1. Space has been provided for you to test a range of values for the factor you have chosen to test.

Running the experiment ▶

ANALYSIS

M Graph your results in the graphing area. Place your time data on the *x*-axis and the number of floating disks on the *y*-axis.

N Use a different-colored pencil to graph the data for each group. Don't forget to either label each resulting curve or provide a key for which group is represented by each color.

CONCLUSION

11. On the basis of your data, determine whether the rate of photosynthesis is affected by the factor you were testing.

_____

12. Was your hypothesis supported or falsified? Use data to support your answer.

_____

_____

_____

13. At what values for the factor (if a range was tested) did photosynthesis happen at the fastest rate?

_____

14. At what values for the factor (if a range was tested) did photosynthesis happen at the slowest rate?

_____

15. What can you conclude regarding the rate of photosynthesis and the factor you were testing?

_____

_____

_____

_____

_____

_____

16. If your results did not support your hypothesis, what factors may have affected your results?

_____

_____

_____

_____

17. What steps would you recommend to improve the experiment you performed?

_____

_____

_____

_____

_____

GOING FURTHER

18. Recall from your textbook that a chemical formula, such as the formula for photosynthesis, is a model, and models need to be tested and, if necessary, modified. What evidence did the leaf disk procedure produce that demonstrated the validity of the photosynthesis equation?

_____

_____

_____

_____

19. In this lab activity you measured something that you could easily observe (floating leaf disks) to make conclusions about a process that you cannot easily observe (photosynthesis). This kind of indirect observation is actually a very common technique in science. Can you think of another process that can be measured indirectly? How is that process measured?

_____

_____

_____

_____

20. When it comes to the need for exercising wise dominion over the earth, food production certainly comes to mind. Why would it be important for farmers to understand what factors influence rates of photosynthesis? How would they apply that understanding to growing crops?

_____

_____

_____

_____

_____

_____

_____

_____

_____

_____

TABLE 1 Number of Floating Leaf Disks

TIME (min)	PRACTICE RUN GROUP (STEP I)	CONTROL GROUP	GROUP 1	GROUP 2	GROUP 3
0					
1					
2					
3					
4					
5					
6					
7					
8					
9					
10					
11					
12					
13					
14					
15					
16					
17					
18					
19					
20					
21					
22					
23					
24					
25					

GRAPHING AREA A

7B LAB

On the Road to Alternative Fuels

Fermentation and Biofuels

In America we can scarcely imagine what life was like before Karl Benz invented the first gas-powered automobile in 1885. Gasoline and diesel fuel, which are used to power most cars, are examples of *fossil fuels*—energy sources derived from the buried remains of plants and animals that have been changed into coal, oil, and natural gas. Because there are no new sources of these fuels being produced today, they are described as *nonrenewable*, meaning that it is possible that one day we could run out of them.

Many people today are engaged in the search for new sources of energy, and *renewable energy* is often their goal. A renewable energy source, as the name suggests, is a source that can be replaced. If a tree is cut down for firewood, for example, a new tree can be planted and grown to replace it. Many scientists and engineers are also looking for *clean energy* sources and *green energy* sources. Clean energy does not produce pollution during its production or use, while green energy sources add little or no greenhouse gases to the atmosphere.

In this lab activity you will investigate some of the issues concerning the production of a particular biofuel—*ethanol*. Ethanol is a kind of alcohol that is frequently used as an additive in gasoline. It isn't extracted from the ground as crude oil is. It's a byproduct of anaerobic cellular respiration, a means of obtaining energy used by many microorganisms, including yeast.

Yeasts use alcoholic fermentation, one form of anaerobic cellular respiration, to extract energy from carbohydrates. In the process, ethanol and carbon dioxide gas are produced as wastes. You should see from this lab activity that we need an inexpensive source of carbohydrates to produce affordable ethanol. In the first part of the activity you will test yeast's ability to ferment different carbohydrate sources, known as *feedstocks*.

? What are some of the economic and societal costs of using fermentation to produce biofuels?

Although many people have invented powered vehicles, Karl Benz's automobile was the first that was powered by gas. The car manufacturer Mercedes-Benz bears his name.

QUESTIONS
- What is fermentation?
- What are biofuels?
- How is fermentation used to produce biofuels?
- Are biofuels worth producing?

Equipment

measuring spoons

graduated cylinder, 50 mL or larger

permanent marker

resealable plastic bags, snack-size (3)

glucose syrup

cornmeal

wood shavings or grass clippings

active dry yeast

warm water (about 40 °C)

goggles

PROCEDURE

Fermenting Carbohydrates

A Label each of the three bags with the feedstock that will be fermented in that bag (glucose syrup, cornmeal, or shavings/clippings).

B In each bag, combine 1 tsp of the appropriate feedstock with 1 tsp of active dry yeast.

C To each bag, add 50 mL of warm water and seal each one after removing as much air as possible. Gently mix the contents of each bag.

D Lay each bag on a flat surface and observe for 15 minutes. If a bag appears to be inflating to the point of bursting, you may release some of the gas.

1. After 15 minutes of observation, rank the bags by the amount of gas produced from most to least.

   _____

2. What is the relationship between the gas produced and the rate of fermentation? Explain.

   _____

   _____

3. On the basis of your results, which of the three do you think would be the best feedstock for producing ethanol? Explain.

   _____

   _____

   _____

The carbohydrates found in the three feedstocks that you tested are different. Glucose syrup contains *glucose*, a simple sugar. Cornmeal is made primarily of *starch*, a long polymer made of many glucose molecules chemically bonded together. Wood shavings, grass clippings, and other plant materials are made mostly of *cellulose*, which is also a long glucose polymer, but the glucose molecules are bonded together in a different manner than those in starch. For cells to process these different carbohydrates, they need customized enzymes that help break down the molecules to release energy and other byproducts.

4. On the basis of your answer to Question 2, which carbohydrate do you think yeast prefers as an energy source?

   _____

5. Why is yeast able to ferment some feedstocks efficiently but not others?

   _____

   _____

   _____

   _____

Research

Now that you've determined which of the three feedstocks ferments best, you may be wondering whether that feedstock is, in fact, the one most used for ethanol production. The answer may surprise you.

E Do some research on the questions and issues surrounding the production and use of ethanol as a biofuel and summarize your findings in a short paper or presentation. Consider the following important topics:

» What are some of the advantages and disadvantages of ethanol compared with those of fossil fuels?

» Who are the primary voices advocating the widespread use of ethanol?

» What are the primary feedstocks being used to produce ethanol?

» Are any other feedstock sources being investigated?

» How much does it cost to produce ethanol?

» How much does ethanol cost for consumers?

» Is ethanol a safe fuel additive?

» Does ethanol yield as much energy per unit of volume as gasoline?

» What are some of the hidden costs of using ethanol on a large scale? (*Hint:* Any feedstock used to make ethanol obviously can't be used for its original purpose.)

F Be prepared to present and discuss your findings when your presentation is due.

CONCLUSION

Clean, renewable energy sounds like a wonderful solution for mankind's energy needs, and perhaps one day it will be. But new technologies often bring along with them some hidden drawbacks. Many hard questions need to be asked and answered before those new technologies gain widespread acceptance. How those questions are answered often depends on one's worldview.

6. Why is there such a push today to develop alternative energies?

_____

_____

_____

_____

7. Are the reasons you stated in Question 6 legitimate justifications for pursuing alternative energies?

_____

_____

_____

_____

8. How does worldview influence a person's view of the cost versus benefits of adopting certain kinds of alternative technology? What role does worldview play in determining the wide-scale adoption of certain kinds of technology?

_____

_____

_____

_____

_____

_____

_____

_____

9. How should a biblical worldview affect our attitudes and opinions about ethanol and other fossil fuel alternatives?

_____

_____

_____

_____

_____

_____

_____

_____

8A LAB

Let's Split

Mitosis and Meiosis

Like the other US states, Wisconsin has a state bird and a state flower—the American robin and the wood violet. But in 2010 Wisconsin lawmakers, in recognition of their state's status as the leading cheese producer in the country, were considering something rather unusual—an official state microbe: the bacterium *Lactococcus lactis*. This microbe loves lactose, the sugar found in milk. When *L. lactis* bacteria are placed in milk, they begin breaking down the milk's lactose for energy, creating lactic acid as a byproduct. Given the abundant supply of lactose in milk, the bacteria begin dividing, making more bacteria that make even more lactic acid. The lactic acid causes the milk to curdle, a process that is the first step in producing several delicious types of cheese, including colby, cheddar, and cottage cheese.

Life as we know it depends on cell division. Cells divide through mitosis to create new cells as old cells die, and many microorganisms, like *L. lactis*, produce new organisms through cell division. In sexual organisms a few cells divide through meiosis to form gametes, which unite to form zygotes that grow and develop into new adults. In this lab activity you will use a microscope to observe cells that have been frozen during the different phases of mitosis and meiosis.

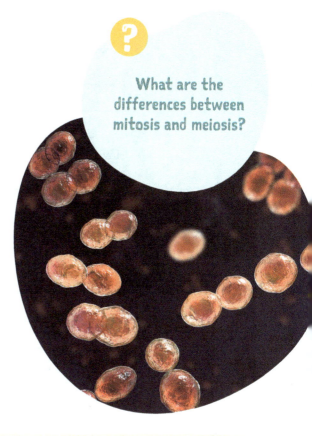

? What are the differences between mitosis and meiosis?

Equipment

microscope
preserved slide of fish embryos prepared for viewing mitosis
preserved slide of fish prepared for viewing meiosis

QUESTIONS

- What are the stages of mitosis?
- What are the stages of meiosis?
- What are the differences between mitosis and meiosis?

PROCEDURE

Mitosis

Mitosis happens rapidly in animal embryos, so if a slide of an embryo is prepared properly, it will reveal all the phases of mitosis.

A Obtain a preserved slide of fish embryos.

1. Indicate and describe the type of sectioning on your slide.

_____

B Observe the slide on high dry or oil immersion power, looking for the various stages of mitosis and interphase. You should find all the phases. Check the phases as you find them.

- ☐ interphase
- ☐ prophase
- ☐ metaphase
- ☐ anaphase
- ☐ telophase
- ☐ daughter cells

2. How did you identify interphase?

_____

_____

_____

3. How did you identify prophase?

_____

_____

_____

4. How did you identify metaphase?

_____

_____

_____

5. How did you identify anaphase?

_____

_____

_____

6. How did you identify telophase?

_____

_____

_____

7. How did you identify the daughter cells?

_____

_____

_____

8. You may have noticed that some of the embryo cells lack chromatin material. Explain the most probable cause for this. (_Hint:_ How were the eggs cut to make this type of slide?)

_____

_____

_____

Use the next page to sketch a series of specimen drawings showing the stages of mitosis in fish embryos. Be sure that you draw typical specimens of the various stages of mitosis.

C Draw one cell in the phase indicated in each of the areas provided. Label
any structures that you can identify.

PROPHASE   METAPHASE

ANAPHASE   TELOPHASE

D Randomly pick twenty cells from the slide and count how many of them are in interphase and in each of the four stages of mitosis.

9. Which stage is more frequently seen than the others?

_____

10. Why do you think this stage is more frequently seen than the others?

_____

_____

Meiosis

While mitosis happens in most of the tissues of a plant or an animal's body, meiosis occurs only in the reproductive organs. So a slide must be made from the reproductive structures in order for meiosis to be seen.

E Obtain a preserved slide of fish meiosis. Observe the slide on high dry or oil immersion power, looking for the various stages of meiosis. You should find all the phases. Check the phases as you find them.

☐ interphase   ☐ prophase II

☐ prophase I   ☐ metaphase II

☐ metaphase I   ☐ anaphase II

☐ anaphase I   ☐ telophase II

☐ telophase I

F Choose one of the stages of meiosis I and draw it and its corresponding stage in meiosis II in the drawing areas below.

DRAWING AREA A   DRAWING AREA B

Comparing Mitosis and Meiosis

Mitosis forms two identical cells; meiosis forms gametes in preparation for sexual reproduction. The processes, while similar, have noticeable differences. Compare mitosis and meiosis by checking the correct choice in the cells in Table 1. All the information you need is in your textbook, but some of the answers may require extra thinking. For telophase assume that the two re-forming nuclei are in separate cells.

TABLE 1 Comparison of Mitosis and Meiosis

TYPE AND PHASE	STAGE CONTAINS	POSITION OF CHROMOSOMES IN THE CELL	NUMBER OF CHROMOSOMES ($2n = 6$)		NUMBER OF CHROMATIDS ($2n = 6$)	
Mitosis Prophase	☐ sister chromatids ☐ daughter chromosomes	☐ moving toward center ☐ located at center ☐ moving from center	☐ 3 ☐ 6	☐ 9 ☐ 12	☐ 6 ☐ 12	☐ 24 ☐ NA
Meiosis Prophase I	☐ sister chromatids ☐ daughter chromosomes	☐ moving toward center ☐ located at center ☐ moving from center	☐ 3 ☐ 6	☐ 9 ☐ 12	☐ 6 ☐ 12	☐ 24 ☐ NA
Mitosis Metaphase	☐ sister chromatids ☐ daughter chromosomes	☐ moving toward center ☐ located at center ☐ moving from center	☐ 3 ☐ 6	☐ 9 ☐ 12	☐ 6 ☐ 12	☐ 24 ☐ NA
Meiosis Metaphase II	☐ sister chromatids ☐ daughter chromosomes	☐ moving toward center ☐ located at center ☐ moving from center	☐ 3 ☐ 6	☐ 9 ☐ 12	☐ 6 ☐ 12	☐ 24 ☐ NA
Mitosis Telophase	☐ sister chromatids ☐ daughter chromosomes	no directional movement	☐ 3 ☐ 6	☐ 9 ☐ 12	☐ 6 ☐ 12	☐ 24 ☐ NA
Meiosis Telophase II	☐ sister chromatids ☐ daughter chromosomes	no directional movement	☐ 3 ☐ 6	☐ 9 ☐ 12	☐ 6 ☐ 12	☐ 24 ☐ NA

11. In which phase of which cellular process would you find tetrads lined up on the equatorial plate?

_____

12. Contrast the purpose of mitosis with the purpose of meiosis in macroorganisms.

_____

_____

_____

8B LAB

The Punnett Square Dance

Inheritance Patterns

In the early twentieth century Gregor Mendel's genetic discoveries were beginning to revolutionize the field of genetics. Geneticists needed a way to predict the possible combinations of alleles in a genetic cross. Reginald Punnett was a young biology professor at Cambridge who had recently published a paper in which he had devised a tool to predict the probability of an organism expressing a particular trait: the Punnett square.

One day he was waiting with mathematician and fellow cricket player G. H. Hardy to play a cricket game that had been delayed by rain. The two professors began to discuss inheritance patterns and gene frequency in a population.

In Chapter 9 you'll learn how Hardy built on Punnett's ideas to understand population genetics. In this lab activity you will use Punnett squares to investigate the different inheritance patterns that can occur in a cross.

How can Punnett squares help scientists learn about inheritance patterns?

Equipment
none

PROCEDURE

Simple Dominance

Some people in the United States can't drink milk—they are lactose intolerant. When they drink milk, they feel bloated, get stomach cramps, and experience discomfort. This might sound like a disease to you, but this condition is perfectly normal in many places in the world. Most people in Africa and Asia are lactose intolerant, but their traditional diets have never included milk anyway. The ability to digest lactose as an adult is found primarily in people of European and Middle Eastern descent.

To digest lactose as an adult, a person must have the ability to produce *lactase*. This ability is an inherited, dominant characteristic. For the exercises that follow, use an uppercase L to represent the dominant allele and a lowercase l to represent the recessive allele.

QUESTIONS
- What are inheritance patterns?
- How can you tell what inheritance pattern a trait has?

A Fill in Table 1 with the correct genotype.

TABLE 1

GENOTYPE	PHENOTYPE
_____	a person who is lactose tolerant
_____	
_____	a person who is lactose intolerant

1. Is there any difference between the phenotypes resulting from the two genotypes that cause a person to be lactose tolerant? Explain.

   _____

   _____

B Jon Forrest, who is homozygous for lactase production, marries Rachel, a lactose intolerant woman. Use this information to fill out the Punnett square in the margin for their offspring and answer Questions 2 and 3.

2. If Jon and Rachel have a daughter who is heterozygous for the lactase gene, will she be able to digest lactose? Explain.

   _____

   _____

3. Could the Forrests have a child who is homozygous recessive for the lactase gene and thus unable to digest lactose? Explain.

   _____

   _____

C Zack, one of the Forrests' sons, marries Ellen, who is known to be heterozygous for the lactase gene. Use this information to fill in the Punnett square in the margin and answer Questions 4–6.

4. Is there a possibility that any of Zack and Ellen's children will be lactose intolerant? If so, circle the genotype(s).

   _____

The *phenotypic ratio* is the ratio of all the possible phenotypes resulting from a cross. The *genotypic ratio* is the ratio of all possible genotypes resulting from a cross. Both are reduced to lowest terms. For example, if a male fruit fly heterozygous for wings (Ww) mates with a homozygous wingless female (ww), the phenotypic ratio of their offspring would be 1 winged : 1 wingless. The genotypic ratio would be 1 Ww : 1 ww.

5. What is the phenotypic ratio for lactose intolerance in Zack and Ellen's children?

   _____

6. What is the genotypic ratio of their children?

   _____

Hugh McTaggart is lactose intolerant, but his wife Amy is lactose tolerant. Amy's mother is lactose intolerant, but Amy's father is lactose tolerant.

7. Is it possible for the McTaggarts to have a child that can digest lactose? Explain.

_____

_____

8. Is it possible for the McTaggarts to have a child that is lactose intolerant? Explain.

_____

_____

9. Explain how you can know Amy's genotype.

_____

_____

_____

10. How might the ability to digest lactose be helpful for a population?

_____

_____

_____

Incomplete Dominance

Though Mendel first discovered the principles of inheritance, we now know that genetics is often much more complex than he imagined. One such complication is *incomplete dominance*.

D When a homozygous red radish plant is crossed with a homozygous white radish plant, purple radishes result.

11. The genotype of a white radish is ____, that of a red radish is ____, and that of a purple radish is ____.

_____

12. In the chart below, write the possible gametes that each type of radish can produce.

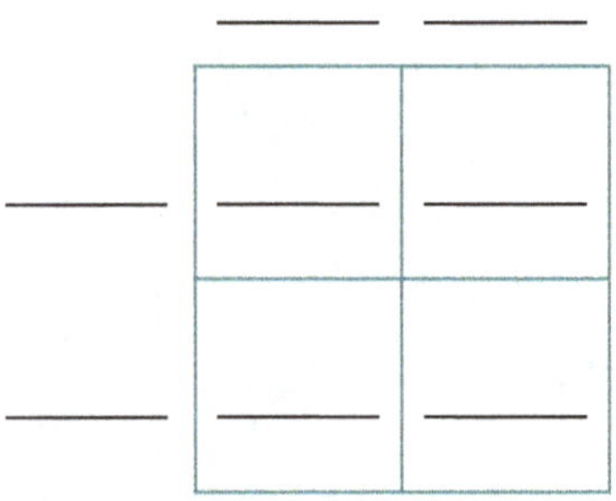

TYPE OF RADISH	POSSIBLE GAMETES
White	_____
Red	_____
Purple	_____

13. If the pollen from a white radish fertilizes the egg of a red radish, what will be the genotypes and the phenotypes of the offspring? Use the Punnett square in the margin to prove your answer.

_____

14. If pollen from a red radish flower fertilizes the egg of a flower on another red radish plant, what will be the genotypes and phenotypes of the offspring?

_____

_____

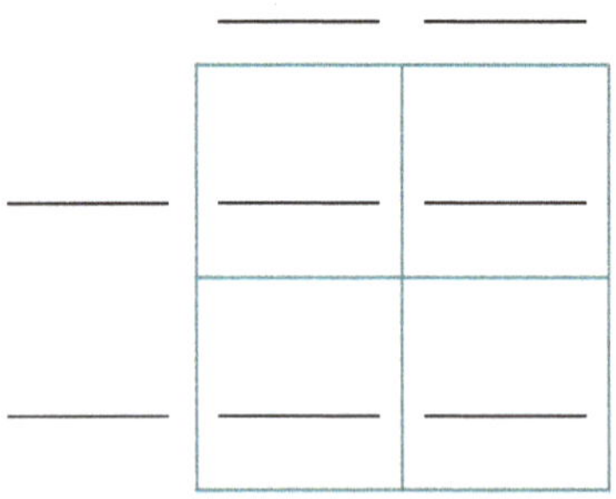

15. If two purple radishes are cross-pollinated, what are the genotypic and phenotypic ratios of the F_1 generation? Prove your answer by making the proper cross on the Punnett square in the margin.

_____

16. If a red radish and a purple radish are cross-pollinated, what will be the phenotypic and genotypic ratios? Complete a Punnett square of the cross in the margin if needed.

_____

_____

17. What would be the genotypic and phenotypic ratios if a white radish and a purple radish were crossed?

_____

_____

Let's consider a similar but more complex problem related to the tails of cats, another trait determined by incomplete dominance.

E The litter resulting from the mating of two short-tailed cats contains two kittens with long tails, six with short tails, and three without tails. In the Punnett square in the margin, diagram a cross that will show the above results.

18. Give a key for the letters you choose to represent the alleles.

_____

_____

_____

19. What are the genotypes of the parents?

_____

20. How does the ratio of the kittens given in the statement compare to the ratio obtained from the Punnett square? Is it close enough for you to be sure that you used the proper genotypes when you diagrammed the cross? Explain.

_____

_____

_____

_____

The fur of some types of mice illustrates incomplete dominance. The fur can be black agouti, normal black, or albino (white). A mouse that is heterozygous is normal black. For Questions 21–25, use F^b to indicate the allele for black agouti fur and F^w to indicate the allele for albino white fur.

21. Determine the black agouti mouse genotype.

_____

22. Determine the normal black mouse genotype.

_____

23. Determine the albino mouse genotype.

_____

F Using the Punnett square in the margin, diagram a cross between a normal black mouse and an albino mouse.

24. What are the genotypic and phenotypic ratios in the F_1 generation?

_____

_____

25. This cross produces a litter of fifteen mice. Eight of the mice are normal black, and seven are albino. Was the expected phenotypic ratio obtained? Explain.

_____

_____

This cow has a roan coat, which resulted from codominance.

Codominance

When a white bull (genotype $C^W C^W$) crosses with a red cow (genotype $C^R C^R$), a roan cow (genotype $C^R C^W$) results.

26. A red bull mates with a red cow. Could their offspring be a roan? Explain.

_____

_____

Two roan cattle mate. Use the Punnett square in the margin to diagram the cross.

27. Give the genotypic and phenotypic ratios for the offspring.

_____

_____

```
_____  _____

     |     |     |
_____|     |     |
     |_____|_____|
     |     |     |
_____|     |     |
     |_____|_____|
```

Cattle typically have only one or two calves at a time, so it's difficult to actually see this phenotypic ratio in real life. However, these principles can be used to solve other real problems.

28. Ned Wright's roan cow is pregnant, and the rancher believes that she has mated with his red bull, but she may have mated with his neighbor's roan bull. The cow gives birth to twins, one roan and one white. Which bull sired the calves? Explain.

_____

_____

29. A farmer with a red cow wants to have roan calves. What color(s) of bull can he cross his cow with in order to have roan calves? If he can get roan calves by crossing his cow with more than one color of bull, which color of bull is most likely to give him roan calves? Draw a Punnett square in the margin if needed.

_____

_____

_____

_____

_____

Multiple Alleles

Sometimes there will be more than one pair of alleles possible at a single locus. Three or more alleles, rather than just two, may be possible.

A gene that controls fur color in rabbits has four different alleles. The wild-type allele, symbolized by R, gives the rabbit a dark color; the exact color depends on several other genes that we will not consider here. The chinchilla allele, denoted by r^{ch}, gives the rabbit a gray look. It is recessive to the wild-type allele but dominant to the other alleles. The Himalayan allele, denoted by r^h, gives a rabbit white fur with dark ears, nose, feet, and tail. It is recessive to the wild-type and chinchilla allele. The albino allele, denoted by r, gives the rabbit an entirely white coat. This allele is recessive to all the other alleles.

G Using the information above, fill in Table 2 with possible genotypes of each phenotype.

TABLE 2

	PHENOTYPE	POSSIBLE GENOTYPE
	wild-type	_____
	chinchilla	_____
	Himalayan	_____
	albino	_____

30. What is the maximum number of color types that could occur in one litter?

_____

_____

31. A male chinchilla rabbit with the genotype of $r^{ch}r^h$ is bred with a female albino rabbit. What will the genotypic and phenotypic ratios be? Use the Punnett square in the margin if you need to.

_____

_____

32. A male wild-type rabbit is crossed with a female Himalayan rabbit, and the resulting litter includes four wild-type rabbits, two Himalayan rabbits, and two albino rabbits. What are the genotypes of the parents?

_____

Polygenic Inheritance

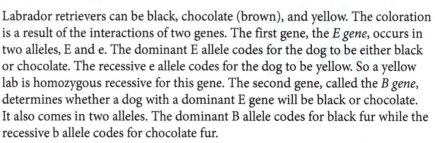

Up to this point we have worked only with traits that are determined by one gene. But this is actually rarely the case. Most traits are the results of interactions between multiple genes.

Labrador retrievers can be black, chocolate (brown), and yellow. The coloration is a result of the interactions of two genes. The first gene, the *E gene*, occurs in two alleles, E and e. The dominant E allele codes for the dog to be either black or chocolate. The recessive e allele codes for the dog to be yellow. So a yellow lab is homozygous recessive for this gene. The second gene, called the *B gene*, determines whether a dog with a dominant E gene will be black or chocolate. It also comes in two alleles. The dominant B allele codes for black fur while the recessive b allele codes for chocolate fur.

33. What is the genotype of a chocolate Labrador puppy whose mother was a yellow lab? How do you know this?

_____

_____

_____

_____

34. A male black lab with a genotype BBEe is crossed with a chocolate female (bbEe). What will the genotypic and phenotypic ratios of the offspring be? Use the Punnett square below.

	_____	_____	_____	_____
_____				
_____				
_____				
_____				

_____

_____

35. A yellow lab is crossed with a black lab. The resulting litter contains three black puppies, two chocolate puppies, and four yellow puppies. The mother of the yellow parent was homozygous dominant for the B gene. What are the genotypes of the two labs?

_____

36. Two yellow labs, both heterozygous for the B gene, are crossed. What are the genotypic and phenotypic ratios of the offspring?

_____

_____

Sex-Linked Traits

In 1911 Thomas Hunt Morgan published a paper suggesting that some genes were carried on chromosomes that we now call sex chromosomes. His discovery of sex-linked traits was based on his studies of inheritance in common fruit flies. The first sex-linked trait he discovered was one that caused the flies' eyes to be white instead of the usual red.

H Fill in the Punnett square below for the cross of a white-eyed male (genotype X^rY) with a wild-type female (X^RX^R).

37. What are the genotypic and phenotypic ratios for the offspring of this cross? Be sure to include the sex of the offspring in the phenotypic ratios.

The fly on the right is a wild-type individual. The fly on the left has white eyes because of a sex-linked mutation.

_____

_____

I One of the female offspring from the previous step mates with a wild-type male. Fill in the Punnett square in the margin for the cross.

38. What are the genotypic and phenotypic ratios for the offspring of this cross?

_____

_____

_____

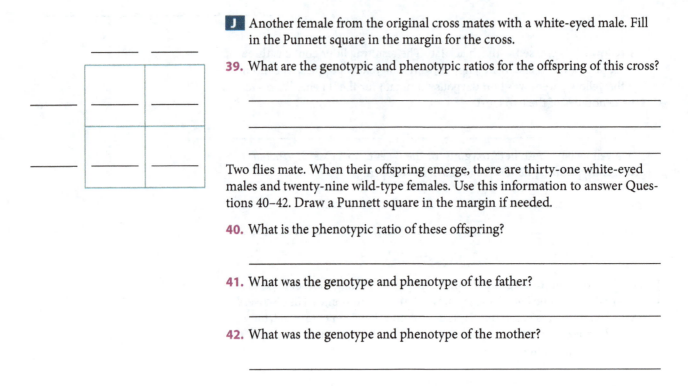

J Another female from the original cross mates with a white-eyed male. Fill in the Punnett square in the margin for the cross.

39. What are the genotypic and phenotypic ratios for the offspring of this cross?

_____

_____

_____

Two flies mate. When their offspring emerge, there are thirty-one white-eyed males and twenty-nine wild-type females. Use this information to answer Questions 40–42. Draw a Punnett square in the margin if needed.

40. What is the phenotypic ratio of these offspring?

_____

41. What was the genotype and phenotype of the father?

_____

42. What was the genotype and phenotype of the mother?

_____

9A LAB

Fix It!

Modeling Genetic Drift

Have you ever worn a warm, comfortable mohair sweater? Mohair is a fiber produced from the wool of Angora goats, believed to be descended from the markhor, a wild goat native to Central Asia. As you can see in the photos on the right, Angora goats look a lot different from their wild cousins! How did this happen? For many centuries humans have been selectively breeding (i.e., artificially selecting) goats domesticated from wild stock. As a result, there are now many varieties of domestic goats raised for meat, milk, and wool or just for getting rid of the poison ivy in your yard.

Charles Darwin and other early evolutionists imagined that natural selection worked much like artificial selection, slowly acting within populations to gradually produce fitter organisms. But later scientists recognized that the gene pools of large populations of organisms tended to remain stable over time. In other words, the change in allele frequency needed to produce new traits for natural selection to act upon was unlikely to occur in large populations. Scientists needed to propose new mechanisms that could produce the necessary change in allele frequencies, also known as genetic drift. In this lab activity you will be simulating genetic drift in a population using beads to represent the alleles for a gene.

?

How do populations acquire new traits?

A wild markhor

A domesticated Angora goat

Equipment

beads, red-colored (100)
beads, white-colored (100)

small container

PROCEDURE

Part 1: Modeling Genetic Drift

A Thoroughly mix together 50 red beads and 50 white beads in a small container. Each color of bead represents an allele for a particular gene. Your container of beads represents a gene pool in which each allele has a frequency of 0.5 (50%).

B Without looking into the container, randomly select 8 pairs of beads. These beads represent the genotypes of Generation 1, that is, the 8 descendants of the original population that will found a new population. Record the frequency of the red and white alleles in this new population in the *Generation 1* row of Table 1 (the allele frequency for Generation 0 has been filled in for you). For example, if 10 of the 16 selected alleles are red, the new frequency would be $10/16 \times 100\% = 62.5\%$ red and 37.5% white.

QUESTIONS

- What is genetic drift?

- How does genetic drift fix an allele within a population?

- Can genetic drift sufficiently explain the production of new phenotypes within a population of organisms?

C Assume that the bead alleles code for color in a simple dominance pattern, with red being the dominant allele. How many of the founders have the red phenotype? How many are white? Record your answers in Table 1.

D The 8 founding individuals that you selected in Step B provide the gene pool for the next generation. Refill your container with 100 beads with the same proportion of red and white beads as for Generation 1. Again, randomly select 8 pairs of beads, representing the genotypes of the next generation (Generation 2). Record the frequency of red and white alleles for this generation in the *Generation 2* row of Table 1.

E How many individuals in Generation 2 are red? How many are white? Record your answers in Table 1.

F Repeat this process, each time recording the allele frequency and number of each phenotype for each generation, until only one kind of allele is present, or until you reach 10 generations.

1. When an allele is eliminated from a population, the remaining allele is said to be *fixed*. Did one of the two alleles become fixed? If so, which one?

_____

2. If an allele became fixed, how many generations did the process take? If neither allele became fixed, what was the lowest frequency reached by one allele?

_____

G Graph the frequency of the red allele in each generation in Graphing Area A. Place the generation number on the *x*-axis and the allele frequency in percent on the *y*-axis.

3. Compare your results with other groups in your class. Did the same allele become fixed for each group? Is this the result that you would expect, given the experimental design? Explain.

_____

_____

_____

_____

_____

_____

4. Did the graphs for the frequency of the red allele look the same for every group? Explain.

_____

_____

_____

_____

_____

5. For this simulation, nothing was said about whether one allele or the other gave an organism a competitive advantage. Suppose the red allele actually gave an organism a competitive advantage. How would that affect the outcome of the experiment?

_____

_____

_____

_____

6. What are some other factors that could affect genetic drift in a population that were not considered in this exercise?

_____

_____

_____

_____

_____

_____

Part 2: Genetic Drift Simulations

Whew! Handling all those beads was a chore, wasn't it? What if you wanted to run the experiment again, but instead of looking at a population of 8 individuals you want to see whether you get the same result for a population of hundreds of organisms? You're going to need a lot more beads!

But don't worry. Scientists and students nowadays have it easier than their counterparts back in the early days of population genetics research. Computers can now simulate things like genetic drift for large populations over many generations. In this section you'll try out one of those simulations.

H Do an internet search using the keywords "genetic drift simulation." You should find several sites that have simple genetic drift simulators. Running the simulation will produce a graph like the one you created in Part 1, showing the trend toward fixing or losing an allele within a population.

I Set the population size to 8 (like the population in Part 1) and the allele frequency to 0.5 and run the simulation. Some simulators allow you to test multiple populations at the same time.

7. What happens to the allele being tested? Is it fixed or lost? Over many generations or just a few?

_____

_____

8. Run the simulation again with the same settings. Do you get the same results? Explain.

_____

_____

_____

9. Now try changing the population size. Run the simulation with populations of 100, 500, and 1000. How is the allele frequency affected?

_____

_____

_____

10. Suppose the initial frequency of a desired allele is less than 0.5. Set the initial frequency to 0.1 and run the simulation for different-sized populations. What happens to the allele frequency?

_____

_____

_____

11. Make a prediction about how the allele frequency will be affected if the initial frequency is greater than 0.5.

_____

_____

_____

12. Now try running the simulation with the initial frequency set at 0.9. How is the allele frequency affected?

_____

_____

_____

13. Using what you have learned from Parts 1 and 2, write a summary of how population size and initial allele frequency affect whether an allele becomes fixed in a population of organisms. Why is this important in a discussion of natural selection?

_____

_____

_____

_____

14. Both creationists and evolutionists recognize that for natural selection to work, new genetic information within a population is necessary. Did the simulations that you used show a gain of new genetic information or a loss of genetic information? Explain. (*Hint:* Remember that the simulations are all based on two or more alleles for a gene. If one increases in frequency, the others must necessarily decrease.)

_____

_____

_____

_____

15. Does selective breeding, such as for the Angora goat, affect genetic information? Explain.

_____

_____

_____

16. How do evolutionists suggest that new information is produced within a population?

17. How does a loss of genetic information fit within a biblical worldview?

_____

_____

_____

_____

_____

_____

_____

_____

_____

_____

Seed banks save heirloom varieties of seeds so that genetic information isn't lost through selective breeding.

GOING FURTHER

So far we have considered only the way that initial allele frequency and population size affect genetic drift. Some simulators also allow you to set values for migration (organisms moving into or leaving a population), bottlenecks (a sudden and dramatic decrease in a population), mutation (the rate at which the alleles being tested mutate), and fitness (an indication of the relative fitness of each possible phenotype that can be produced by combinations of the tested alleles).

18. Try experimenting with the other simulation settings and write a brief summary of what you discover.

_____

_____

_____

_____

_____

TABLE 1 Frequency of Red and White Alleles in Bead Populations

GENERATION	RED ALLELE FREQUENCY (%)	WHITE ALLELE FREQUENCY (%)	NUMBER OF RED PHENOTYPE (RR or Rr)	NUMBER OF WHITE PHENOTYPE (rr)
0	50	50		
1				
2				
3				
4				
5				
6				
7				
8				
9				
10				

9B LAB

Whodunit?

DNA Fingerprinting

A crime has occurred, and the culprit has been caught and punished. Justice has been served! Or has it?

DNA fingerprinting is an incredibly useful tool in forensics. It is used to analyze evidence in criminal investigations, resolve paternity cases, and identify victims of crimes or disasters. The idea behind this technique is that each individual has unique DNA sequences. Though there are several DNA sequences that all people and even some animals share, everyone has some unique DNA sequences in noncoding regions. These sequences can show the difference between individuals, just like fingerprints from a person's finger. What might surprise you is that DNA fingerprinting has shown a large number of people, already convicted of a crime, to be innocent rather than guilty.

The process of DNA fingerprinting involves a series of techniques that produce a pattern of DNA that can be analyzed. This pattern looks like a bar code. To get this pattern, DNA must be isolated and purified in a way that is similar to the technique that you learned in Lab 6B. In this lab activity you will simulate the steps of DNA fingerprinting to analyze some "DNA samples."

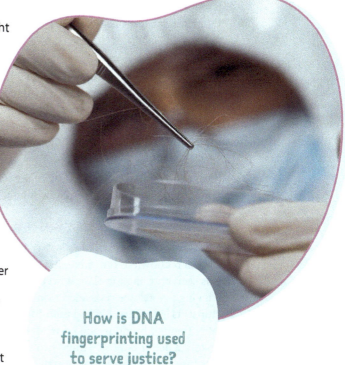

How is DNA fingerprinting used to serve justice?

?

Equipment

QUESTIONS
- What is DNA fingerprinting?
- How is DNA fingerprinting done?
- How is DNA evidence used to help solve crimes?

THE CRIME

The day was going to be perfect! The sophomore class had developed a flawless recipe for Super Chocolate Brownies. They were entering their recipe, along with samples, into the school bakeoff competition, the tiebreaking event in this year's school spirit week. But when Mr. Sokolata, the sophomore class advisor, entered his classroom after lunch, he was shocked to find that the freshly baked brownies, which he had left on his desk to cool, had been mostly eaten, and the recipe card was missing! Who could have done such a thing?

No one saw what happened, but on the basis of an abundance of circumstantial evidence, a student named Chip Lee was charged with theft that very afternoon and expelled from school. The case didn't sit well with Mrs. Gumschuh, the biology teacher. She decided to look into the case. Happily, Mr. Sokolata had no afternoon classes in his room, and the crime scene was still mostly undisturbed when Mrs. Gumschuh stopped by after school.

THE EVIDENCE

Here's what she found. The baking pan, a few partially eaten brownies still in it, remained on the desk where Mr. Sokolata had left it. At a student desk near the pan of brownies was a senior class T-shirt with some hairs on the collar. Mrs. Gumschuh learned from Mr. Stern, the dean of students, that no sign had been found of forced entry into the classroom from outside the building. Mrs. Gumschuh collected samples of the hair from the T-shirt and swabbed the brownie pan for saliva to test for DNA. She also collected cheek swabs from each of the four original suspects. Read on.

Suspect Data

After interviewing students, Mr. Stern had narrowed the list of suspects down to four individuals. Circumstantial evidence pointed to the following persons.

SUSPECT 1: I. K. "BEN" ORDELIJK, CUSTODIAN

Mr. Ordelijk is in charge of cleaning the classrooms and has keys to every room at the school, so he definitely had access to the brownies, even if the door hadn't been open. He's also been telling other staff members for weeks that he's been craving sugar since starting a new low-carb diet.

SUSPECT 2: CHIP LEE, STUDENT

Chip has been an unwilling visitor to Mr. Stern's office on several occasions and has been known to try to talk his way out of consequences for his actions. Chip was seen entering the classroom by Mr. Ordelijk, though he claims he was retrieving a notebook. His addiction to brownies (especially those with chocolate chips!) is legendary. He had mentioned to some students earlier that Mr. Sokolata often left his door open during lunch and had reportedly quipped, "Boy, Mr. Sokolata better not leave those brownies where I can get to them!" Chip usually eats lunch with the same group of students every day, but on the day of the crime he arrived late and had dark-colored crumbs on his clothing.

SUSPECT 3: BROCK LEE, STUDENT

Brock is Chip's twin brother and Mr. Sokolata's classroom aid. He stops by Mr. Sokolata's room every day at lunch to restock supplies, but it's common knowledge that he prefers doughnuts to brownies.

SUSPECT 4: PATSY FAHLGUY, SENIOR CLASS PRESIDENT

Patsy had vowed that the senior class would win the spirit week competition at any cost. She had all the seniors wear their class T-shirts to show team spirit.

DNA Analysis

Mrs. Gumschuh isolated DNA from the hair sample, saliva sample, and cheek cell samples from each of the suspects. Next, she cut the DNA with a restriction enzyme. The restriction enzyme Mrs. Gumschuh used always cuts at the sequence ATTA between the two Ts.

1. The restriction enzyme used in this step cut DNA at the same specific site every time. Why is this essential for the procedure?

   _____

   _____

   _____

A DNA sequences for each of the samples are listed below. Draw a line through each cut site in each DNA sample. The saliva DNA sample has been done for you as an example.

SALIVA DNA SAMPLE:

A T G T A G A C T G G A C C A T A T|T A C

G A T|T A G G C A C T C A T|T A G C C G T

A C A G T A C T C A C C

HAIR DNA SAMPLE:

A T C T C G T G A C A T T A C C T T G T A T

C G A T T A G C A A T T A A G G A T C C T

G C A G T A G C A C C

SUSPECT 1 DNA SAMPLE:

A T T A A C G G G T A T C T T C G G A T T A

C G G A G A C T A A G T G C C T A G A T T

A C G A A G C T A C C

SUSPECT 2 DNA SAMPLE:

A T C A G C A T G T G T T C A A T T A G C C

G A G A T T A A G G C C A C T G G A G T A

C T A C G G C C A C C

SUSPECT 3 DNA SAMPLE:

A T G T A G A C T G G A C C A T A T T A C

G A T T A G G C A C T C A T T A G C C G T

A C A G T A C T C A C C

SUSPECT 4 DNA SAMPLE:

A T C T C G T G A C A T T A C C T T G T A T

C G A T T A G C A A T T A A G G A T C C T

G C A G T A G C A C C

2. What do all the DNA sequences have in common? Explain this similarity.

_____

_____

_____

Now you will need to count the number of base pairs in the fragments (created by the cut sites) of each DNA sample. For example, the saliva sample has four fragments, with the first fragment having 18 base pairs (bp), then 6 bp, 11 bp, and 19 bp.

3. Count the fragment lengths in the hair DNA sample and each suspect DNA sample.

Hair: _____

Suspect 1: _____

Suspect 2: _____

Suspect 3: _____

Suspect 4: _____

The DNA fragments are sorted by loading each sample of DNA into the well of a gel. An electric current moves the negatively charged DNA fragments toward the positive end of the gel, sorting them by size with the smaller pieces moving the farthest. This creates a unique pattern, or DNA fingerprint.

B Use Table 1 to sort the DNA fragments by drawing a dash in the gel box that corresponds to the fragment length. The saliva sample is done for you as an example.

Next, Mrs. Gumschuh stained the DNA gel with a fluorescent dye.

4. Why do you think this step is essential in the DNA fingerprinting sequence?

_____

_____

EVALUATING THE EVIDENCE

Now let's analyze the evidence to find out whodunit!

5. Whom does the hair sample belong to?

_____

6. Whom does the saliva sample belong to?

_____

7. Who committed the crime?

_____

8. How do you know that this person committed the crime?

_____

_____

_____

9. Using the evidence above, explain how DNA fingerprinting can be used to clear a person as a suspect even though circumstantial evidence places that person at the scene of the crime.

_____

_____

_____

_____

_____

10. Why were Chip's and Brock's DNA fingerprints not identical even though they are twins?

_____

_____

_____

11. What are some limitations to using this technique as evidence in a trial?

_____

_____

_____

_____

_____

_____

_____

_____

_____

_____

GOING FURTHER

Eating a pan of brownies and getting expelled is not the worst thing that could happen to a person. Many people have been convicted of far worse crimes and sent to prison, often for many years, on the basis of less than completely convincing evidence. The Innocence Project is a nonprofit organization that seeks to exonerate people wrongfully convicted of crimes. As of 2021 they have gotten nearly 200 convictions overturned; some of the exonerated were even on death row. Do an internet search using the keywords "Innocence Project" and answer the following questions.

12. What is the primary way in which the Innocence Project exonerates people wrongfully convicted of crimes?

_____

_____

13. The Innocence Project is not always successful in getting convictions overturned. Why not?

_____

_____

14. How does the work of the Innocence Project fit with a biblical worldview?

_____

_____

_____

_____

TABLE 1 Electrophoresis Gel Box

	SALIVA	HAIR	SUSPECT 1	SUSPECT 2	SUSPECT 3	SUSPECT 4
30						
29						
28						
27						
26						
25						
24						
23						
22						
21						
20						
19	▬▬▬▬					
18	▬▬▬▬					
17						
16						
15						
14						
13						
12						
11	▬▬▬▬					
10						
9						
8						
7						
6	▬▬▬▬					
5						
4						
3						
2						
1						

10A LAB

In Darwin's Own Words

Examining On the Origin of Species

You have been learning about Charles Darwin's ideas and how they were shaped by his culture and times. Let's look at the introduction to *On the Origin of Species*. You will read Darwin's work and analyze his words by answering some questions. As you read, look for Darwin's three themes: variation, the struggle for survival, and natural selection in species. Be ready to evaluate his ideas from a biblical perspective.

What is *On the Origin of Species* all about?

Equipment

QUESTIONS
- What are Darwin's key points?
- What does Darwin say that is compatible with a biblical worldview?
- What does Darwin say that is incompatible with a biblical worldview?

INTRODUCTION FROM ON THE ORIGIN OF SPECIES

When on board H.M.S. 'Beagle,' as naturalist, I was much struck with certain facts in the distribution of the inhabitants of South America, and in the geological relations of the present to the past inhabitants of that continent. These facts seemed to me to throw some light on the origin of species—that mystery of mysteries, as it has been called by one of our greatest philosophers. On my return home, it occurred to me, in 1837, that something might perhaps be made out on this question by patiently accumulating and reflecting on all sorts of facts which could possibly have any bearing on it. After five years' work I allowed myself to speculate on the subject, and drew up some short notes; these I enlarged in 1844 into a sketch of the conclusions, which then seemed to me probable: from that period to the present day I have steadily pursued the same object. I hope that I may be excused for entering on these personal details, as I give them to show that I have not been hasty in coming to a decision.

My work is now nearly finished; but as it will take me two or three more years to complete it, and as my health is far from strong, I have been urged to publish this Abstract. I have more especially been induced to do this, as Mr. Wallace, who is now studying the natural history of the Malay Archipelago, has arrived at almost exactly the same general conclusions that I have on the origin of species. Last year he sent to me a memoir on this subject, with a request that I would forward it to Sir Charles Lyell, who sent it to the Linnean Society, and it is published in the third volume of the journal of that Society. Sir C. Lyell and Dr. Hooker, who both knew of my work—the latter having read my sketch of 1844—honoured me by thinking it advisable to publish, with Mr. Wallace's excellent memoir, some brief extracts from my manuscripts.

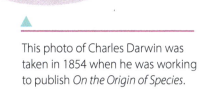

This photo of Charles Darwin was taken in 1854 when he was working to publish *On the Origin of Species*.

This Abstract, which I now publish, must necessarily be imperfect. I cannot here give references and authorities for my several statements; and I must trust to the reader reposing some confidence in my accuracy. No doubt errors will have crept in, though I hope I have always been cautious in trusting to good authorities alone. I can here give only the general conclusions at which I have arrived, with a few facts in illustration, but which, I hope, in most cases will suffice. No one can feel more sensible than I do of the necessity of hereafter publishing in detail all the facts, with references, on which my conclusions have been grounded; and I hope in a future work to do this. For I am well aware that scarcely a single point is discussed in this volume on which facts cannot be adduced, often apparently leading to conclusions directly opposite to those at which I have arrived. A fair result can be obtained only by fully stating and balancing the facts and arguments on both sides of each question; and this cannot possibly be here done.

I much regret that want of space prevents my having the satisfaction of acknowledging the generous assistance which I have received from very many naturalists, some of them personally unknown to me. I cannot, however, let this opportunity pass without expressing my deep obligations to Dr. Hooker, who for the last fifteen years has aided me in every possible way by his large stores of knowledge and his excellent judgment.

In considering the origin of species, it is quite conceivable that a naturalist, reflecting on the mutual affinities of organic beings, on their embryological relations, their geographical distribution, geological succession, and other such facts, might come to the conclusion that each species had not been independently created, but had descended, like varieties, from other species. Nevertheless, such a conclusion, even if well founded, would be unsatisfactory, until it could be shown how the innumerable species inhabiting this world have been modified so as to acquire that perfection of structure and coadaptation which most justly excites our admiration. Naturalists continually refer to external conditions, such as climate, food, etc., as the only possible cause of variation. In one very limited sense, as we shall hereafter see, this may be true; but it is preposterous to attribute to mere external conditions, the structure, for instance, of the woodpecker, with its feet, tail, beak, and tongue, so admirably adapted to catch insects under the bark of trees. In the case of the mistletoe, which draws its nourishment from certain trees, which has seeds that must be transported by certain birds, and which has flowers with separate sexes absolutely requiring the agency of certain insects to bring pollen from one flower to the other, it is equally preposterous to account for the structure of this parasite, with its relations to several distinct organic beings, by the effects of external conditions, or of habit, or of the volition of the plant itself.

The author of the *Vestiges of Creation* would, I presume, say that, after a certain unknown number of generations, some bird had given birth to a woodpecker, and some plant to the mistletoe, and that these had been produced perfect as we now see them; but this assumption seems to me to be no explanation, for it leaves the case of the coadaptations of organic beings to each other and to their physical conditions of life, untouched and unexplained.

▲

Darwin used the relationship of mistletoe, a parasitic plant, to the birds that spread its seeds to demonstrate that ecological relationships are the product of natural selection.

It is, therefore, of the highest importance to gain a clear insight into the means of modification and coadaptation. At the commencement of my observations it seemed to me probable that a careful study of domesticated animals and of cultivated plants would offer the best chance of making out this obscure problem. Nor have I been disappointed; in this and in all other perplexing cases I have invariably found that our knowledge, imperfect though it be, of variation under domestication, afforded the best and safest clue. I may venture to express my conviction of the high value of such studies, although they have been very commonly neglected by naturalists.

From these considerations, I shall devote the first chapter of this Abstract to Variation under Domestication. We shall thus see that a large amount of hereditary modification is at least possible, and, what is equally or more important, we shall see how great is the power of man in accumulating by his Selection successive slight variations. I will then pass on to the variability of species in a state of nature; but I shall, unfortunately, be compelled to treat this subject far too briefly, as it can be treated properly only by giving long catalogues of facts. We shall, however, be enabled to discuss what circumstances are most favourable to variation. In the next chapter the Struggle for Existence among all organic beings throughout the world, which inevitably follows from their high geometrical powers of increase, will be treated of. This is the doctrine of Malthus, applied to the whole animal and vegetable kingdoms. As many more individuals of each species are born than can possibly survive; and as, consequently, there is a frequently recurring struggle for existence, it follows that any being, if it vary however slightly in any manner profitable to itself, under the complex and sometimes varying conditions of life, will have a better chance of surviving, and thus be naturally selected. From the strong principle of inheritance, any selected variety will tend to propagate its new and modified form.

Darwin observed much variation in domesticated species. Among his favorite animals to observe were pigeons. He even began to breed pigeons himself around the time *On the Origin of Species* was published.

This fundamental subject of Natural Selection will be treated at some length in the fourth chapter; and we shall then see how Natural Selection almost inevitably causes much Extinction of the less improved forms of life and induces what I have called Divergence of Character. In the next chapter I shall discuss the complex and little known laws of variation and of correlation of growth. In the four succeeding chapters, the most apparent and gravest difficulties on the theory will be given: namely, first, the difficulties of transitions, or understanding how a simple being or a simple organ can be changed and perfected into a highly developed being or elaborately constructed organ; secondly the subject of Instinct, or the mental powers of animals; thirdly, Hybridism, or the infertility of species and the fertility of varieties when intercrossed; and fourthly, the imperfection of the Geological Record. In the next chapter I shall consider the geological succession of organic beings throughout time; in the eleventh and twelfth, their geographical distribution throughout space; in the thirteenth, their classification or mutual affinities, both when mature and in an embryonic condition. In the last chapter I shall give a brief recapitulation of the whole work, and a few concluding remarks.

No one ought to feel surprise at much remaining as yet unexplained in regard to the origin of species and varieties, if he makes due allowance for our profound ignorance in regard to the mutual relations of all the beings which live around us. Who can explain why one species ranges widely and is very numerous, and why another allied species has a narrow range and is rare? Yet these relations are of the highest importance, for they determine the present welfare, and, as I believe, the future success and modification of every inhabitant of this world. Still less do we know of the mutual relations of the innumerable inhabitants of the world during the many past geological epochs in its history. Although much remains obscure, and will long remain obscure, I can entertain no doubt, after the most deliberate study and dispassionate judgment of which I am capable, that the view which most naturalists entertain, and which I formerly entertained—namely, that each species has been independently created—is erroneous. I am fully convinced that species are not immutable; but that those belonging to what are called the same genera are lineal descendants of some other and generally extinct species, in the same manner as the acknowledged varieties of any one species are the descendants of that species. Furthermore, I am convinced that Natural Selection has been the main but not exclusive means of modification.

Charles Darwin.

ANALYZING DARWIN'S WORDS

1. On what did Darwin base his conclusions about where species come from?

   _____

   _____

   _____

   _____

2. Name three of the other scientists Darwin mentions and describe how they influenced his work.

   _____

   _____

   _____

   _____

   _____

   _____

3. What does Darwin plan to discuss in Chapter 1? How does this apply to the origin of species?

   _____

   _____

   _____

   _____

4. What does Darwin plan to discuss in Chapter 2? Who inspired his ideas in this chapter?

_____

_____

_____

_____

5. What does Darwin plan to discuss in Chapter 4?

_____

6. How does Darwin say natural selection affects a population?

_____

_____

_____

7. What is Darwin's position on fixity of species?

_____

8. Name at least three things that you can agree with Darwin about.

_____

_____

_____

_____

9. Name one thing that you disagree with Darwin about.

_____

_____

_____

_____

10. Name one weakness of Darwin's work.

_____

_____

_____

11. Name one strength of Darwin's work.

_____

_____

_____

12. Darwin and Chambers believed in God, but they viewed Him as someone who worked in the world through the laws of science without being personally involved. Comment on this view of God from a biblical perspective.

_____

_____

_____

_____

_____

_____

13. What was Darwin's source of reliable truth for his ideas about the origin of life?

_____

_____

_____

14. Analyze the authority for Darwin's ideas from a biblical perspective.

_____

_____

_____

GOING FURTHER

A Consider reading other chapters from this work that interest you and share your thoughts with someone else.

B Another of Darwin's works, *The Descent of Man*, was published in 1871. In this book Darwin takes his ideas on the evolution of animals and applies them to mankind. This book contains some very controversial ideas related to social Darwinism, including the notion that weak individuals in the human population should not be allowed to reproduce so that they don't weaken the human race. Social Darwinism began to affect public thinking in the 1870s. The moral implications of social Darwinism are completely contrary to the central principles of Scripture. Read a selection from *The Descent of Man* and share your thoughts with someone else.

10B LAB

Worldview Sleuthing

Evaluating Worldview in Popular Science Literature

Do you or your teacher have a subscription to *National Geographic*? *Scientific American*? *Science News*? *Popular Science*? These periodicals, like many others, can give you interesting information in ways that are easy to understand. Television programs like *NOVA* and *Nature* serve a similar purpose.

Have you ever read something in these periodicals or seen something on these programs that made you uncomfortable? Probably. But why? We feel uncomfortable when someone says or writes something that contradicts what we believe, especially when it is one of our core beliefs. In this lab activity you will read three snippets from popular science magazines and analyze them from a biblical perspective.

Should I believe everything I read?

Equipment

current issues of science periodicals or digital copies of online articles

QUESTIONS

- What controversial issues do people write about in popular science periodicals?

- How can I identify a science writer's worldview?

- How can I tell what to believe about what I read?

PROCEDURE

Evolution is not just a theory; it forms a philosophical framework that shapes how people view life. This extends to ideas about extraterrestrial life and conservation here on planet Earth. Evolution even affects people's ideas about culture and technology and the future of mankind. You'll look at excerpts from *National Geographic* and *Science News* on these topics.

Extraterrestrial Life

The first quote comes from an article about the history of the search for extraterrestrial life. Recent efforts to find alien lifeforms focus on exoplanets in the "Goldilocks" zone of their host star—planets that have just the right conditions for life to evolve.

"The universe is vast and old, so advanced civilizations should have matured enough by now to send emissaries to Earth. Yet none have. Fermi suspected that it wasn't feasible or that aliens didn't think visiting Earth was worth the trouble. Others concluded that they simply don't exist. Recent investigations indicate that harsh environments may snuff out nascent life long before it evolves the intelligence necessary for sending messages or traveling through space."[1]

An organization called SETI (Search for Extraterrestrial Intelligence) uses a group of telescopes called the Allen Array to survey the universe for signs of extraterrestrial life.

---

[1] "To An Ancient Question, No Reply" by Tom Siegfried, *Science News*, April 30, 2016, pp. 24–25

1. What words or phrases in the article suggest that the writer of this article is not writing from a biblical worldview?

_____

_____

_____

_____

_____

2. What do scientists assume about how life should evolve on other planets? Is that a good assumption? Explain.

_____

_____

_____

_____

_____

3. How do evolutionists relate intelligence and communication to life?

_____

_____

_____

Bison grazing in Yellowstone National Park

Conservation

The second excerpt comes from an issue of *National Geographic* dedicated to the history and conservation of Yellowstone National Park. The quote below refers to a conversation between the writer and Dave Hallac, chief of the Yellowstone Center for Resources, which oversees programs that maintain the ecological health of the park.

"Dave Hallac stood in his office at the Yellowstone Center for Resources, the park body charged with science and resource management, in a rambling old clapboard building amid the formidable stone structures of Mammoth Hot Springs. Hallac ticked through a list of inter-related concerns, nagging issues in Yellowstone familiar to us both: bison management, elk migration, grizzly bear conservation, private land development in the region surrounding the park, human population growth driving that development, invasive species and their impacts on native species, water use, climate change, and finally the overarching problem that exacerbates all these others—an absence of coordinated, transboundary management. 'We go around telling everybody this is the most intact ecosystem in the lower 48,' Hallac said. Well, if it's that important, that special, it's time for us to do a lot better when it comes to protecting it."[2]

[2] "Living With the Wild" by David Quammen, *National Geographic*, May 2016, p. 133

4. Is there anything in this article that is at odds with a biblical worldview? Explain.

_____

_____

_____

_____

_____

5. Most of the conservationists at work in Yellowstone are evolutionists. Comment on the ability of evolutionists to do good science from a biblical perspective.

_____

_____

_____

_____

_____

6. Is a growing human population a good thing or a bad thing from a biblical worldview? Why?

_____

_____

_____

7. Conservationists at Yellowstone National Park are working to protect the park but still keep it accessible to a growing number of visitors. How does this show the interaction between the two parts of the Creation Mandate?

_____

_____

_____

8. Why do evolutionists value conservation? Analyze their reasons in light of their other beliefs.

_____

_____

_____

_____

_____

_____

_____

A final quote comes from an article about peoples' near-death experiences. Death is a common human experience, and evolution relies on death for populations to change over time. Humans are the most evolved species, according to this worldview, so in the future they should be able to find a way to fight against the effects of death.

"Death is 'a process, not a moment,' writes critical-care physician Sam Parnia in his book *Erasing Death*. It's a whole-body stroke, in which the heart stops beating but the organs don't die immediately. In fact, he writes, they might hang on intact for quite a while, which means that 'for a significant period of time after death, death is in fact fully reversible.'

"How can death, the very essence of forever, be reversible? What is the nature of consciousness during that transition through the gray zone? A growing number of scientists are wrestling with such vexing questions.

"In Seattle biologist Mark Roth experiments with putting animals into a chemically induced suspended animation, mixing up solutions to lower heartbeat and metabolism to near-hibernation levels. His goal is to make human patients who are having heart attacks 'a little bit immortal' until they can get past the medical crisis that brought them to the brink of death."[3]

[3] "The Crossing" by Robin Marantz Henig, *National Geographic*, April 2016, p. 36

Cryonics uses deep freezing to store a person already declared legally dead with the hope that the person can be brought back to life sometime in the future.

9. Is it wrong to try to extend people's lives through medical intervention? Explain.

_____

_____

_____

_____

_____

10. If death is such a big player in evolution, why do you think that people without a biblical worldview fear it and try to work against it? Why not let nature do its work?

_____

_____

_____

_____

_____

11. How might scientists and doctors operate within a biblical worldview even though they might not be Christians?

_____

_____

_____

_____

12. From a biblical perspective, will people and technology ever advance to the degree that death can be completely overcome? Explain.

_____

_____

_____

_____

GOING FURTHER

Find some articles of your own to do some worldview sleuthing. You can check *National Geographic*, *Smithsonian*, *Popular Mechanics*, and *Scientific American*, looking either at physical copies or exploring online.

13. What topic did you read about?

_____

14. What is the viewpoint of the writer on this topic?

_____

_____

_____

_____

_____

_____

15. Analyze this viewpoint from a biblical worldview.

_____

_____

_____

_____

_____

_____

11A LAB

The Key Concept

Using Dichotomous Keys

Edmund and Tenzing were out backpacking in the Colorado foothills. They both enjoy the taste of wild onions that grow in the area, such as Geyer's onion (*Allium geyeri*). They also found another plant that looked very much like an onion but lacked the strong wild-onion smell. They decided to avoid that plant since they know that most species in the genus Allium smell like onions. The two adventurers had performed a simplified version of classification, and though they might not have realized it, their lives had depended on accurately identifying the unfamiliar plant as a member of the genus Allium or not. The unknown onion look-alike happened to be a meadow death camas (*Toxicoscordion venenosum*), a plant whose common name correctly suggests that it is not a suitable trailside treat!

Like our intrepid hikers, biologists often need to identify unknown organisms, and a commonly used tool for the task is a dichotomous key. It's a series of paired statements or characteristics about the specimen being identified. Only one statement from each pair can be true, and selecting the correct statement either identifies the organism or refers the user to another pair of statements to consider. Today you will use a simplified dichotomous key to sort out some finned friends.

Tasty!

Why is being able to identify organisms important?

Deadly!

Equipment

Key for Selected Fishes of North America (pages 119–20)

PROCEDURE

Use the key on pages 119–20 to identify the fish on pages 112–18.

A Start by reading the first pair of statements, then decide which of the two statements describes your specimen. The statement you choose will either give you the identity of your fish or direct you to another pair of statements.

B Record each number from the key that you use to identify each organism in Table 1.

C When you have finally identified your fish, record its scientific and common names in Table 1.

D Repeat Steps A through C for additional specimens. Specimen 1 has been completed for you.

QUESTIONS

- What is a dichotomous key?
- How does a dichotomous key work?
- How can a dichotomous key help identify an unknown organism?

TABLE 1

Specimen Number	Numbers from Key	Common Name	Scientific Name
1	1, 2, 4, 5, 8, 9, 10	rainbow trout	*Oncorhynchus mykiss*
2			
3			
4			
5			
6			
7			
8			
9			
10			
11			
12			
13			
14			
15			
16			
17			
18			
19			
20			

Specimen Number	Numbers from Key	Common Name	Scientific Name
21			
22			
23			
24			
25			
26			
27			
28			

GOING FURTHER

1. Like most science tools, a dichotomous key has some limits to its usefulness. What are some things you can think of that might hinder the effectiveness of a dichotomous key as a tool for identifying organisms?

_____

_____

_____

_____

_____

2. One of the first tasks that God assigned to Adam in Eden was the giving of names to all the animals that God brought to him (Gen. 2:19–20). On the basis of what you learned about speciation in Chapter 10, do you think that Adam had a tougher job than modern biologists? Or was his easier? Explain.

_____

_____

_____

_____

_____

_____

_____

5

7

6

8

10

12

9

11

13

15

14

16

18

20

17

19

25

26

27

28

Key to Selected Fishes of North America

1. (a) Body is covered with bony scutes . . . . Atlantic sturgeon (*Acipenser oxyrinchus*)

(b) Body is covered with scales or is scaleless . . . . go to 2

2. (a) Has a heterocercal caudal fin . . . . go to 3

(b) Has a homocercal caudal fin . . . . go to 4

3. (a) Has a short snout and long dorsal fin . . . . bowfin (*Amia calva*)

(b) Has a long snout and short dorsal fin . . . . longnose gar (*Lepisosteus osseus*)

4. (a) Has an adipose fin . . . . go to 5

(b) Does not have an adipose fin . . . . go to 11

5. (a) Has barbels . . . . go to 6

(b) Does not have barbels . . . . go to 8

6. (a) Caudal fin is rounded . . . . brindled madtom (*Noturus miurus*)

(b) Caudal fin is forked . . . . go to 7

7. (a) Caudal fin is deeply forked . . . . channel catfish (*Ictalurus punctatus*)

(b) Caudal fin is slightly forked . . . . black bullhead (*Ameiurus melas*)

8. (a) Caudal fin is deeply forked . . . . lake whitefish (*Coregonus clupeaformis*)

(b) Caudal fin is slightly or moderately forked . . . . go to 9

9. (a) Has a prominent red "slash" mark under jaw . . . . . . Bonneville cutthroat trout (*Oncorhynchus clarkii utah*)

(b) Has no red "slash" mark under jaw . . . . go to 10

10. (a) Has black spots on pectoral, pelvic, and anal fins . . . . rainbow trout (*Oncorhynchus mykiss*)

(b) Has pectoral, pelvic, and anal fins red with no spots . . . . brook trout (*Salvelinus fontinalis*)

11. (a) Has one dorsal fin with only soft rays . . . . go to 12

(b) Has one or two dorsal fins with spiny and soft rays . . . . go to 18

12. (a) Leading edge of dorsal fin is at or near midpoint of body . . . . go to 13

(b) Dorsal fin is set far back on body . . . . go to 16

13. (a) Mouth is terminal . . . . American shad (*Alosa sapidissima*)

(b) Mouth is subterminal or inferior . . . . go to 14

14. (a) Has barbels . . . . common carp (*Cyprinus carpio*)

(b) Does not have barbels . . . . go to 15

15. (a) First rays of dorsal fin are nearly equal in length . . . . silver redhorse (*Moxostoma anisurum*)

(b) First rays of dorsal fin are greatly elongated . . . . river carpsucker (*Carpiodes carpio*)

16. (a) Has no spots on fins . . . . redfin pickerel (*Esox americanus*)

(b) Has spots on fins . . . . go to 17

17. (a) Has irregular vertical barring on sides . . . . muskellunge (*Esox masquinongy*)

(b) Has light spots on olive green sides . . . . northern pike (*Esox lucius*)

18. (a) Spiny dorsal fin and soft dorsal fin are fused . . . . go to 19

(b) Spiny dorsal fin and soft dorsal fin are separate or nearly separate . . . . go to 23

19. (a) Caudal fin is rounded with no fork . . . . blue-spotted sunfish (*Enneacanthus gloriosus*)

    (b) Caudal fin is forked . . . . go to 20

20. (a) Fins are reddish-orange with no spots . . . . go to 21

    (b) Fins are heavily spotted . . . . go to 22

21. (a) Has white vermiculated markings on operculum . . . . pumpkinseed (*Lepomis gibbosus*)

    (b) Has black, greatly lengthened anterior margin of operculum . . . . redbreast sunfish (*Lepomis auritus*)

22. (a) Sides are silvery white with black vertical barring . . . . white crappie (*Pomoxis annularis*)

    (b) Sides are white with heavy black spotting . . . . black crappie (*Pomoxis nigromaculatus*)

23. (a) Sides are yellow with eight or nine prominent, black vertical bars . . . . yellow perch (*Perca flavescens*)

    (b) Sides are silver or olive . . . . go to 24

24. (a) Sides are olive . . . . go to 25

    (b) Sides are silver . . . . go to 26

25. (a) Posterior margin of mouth extends past eye . . . . largemouth bass (*Micropterus salmoides*)

    (b) Posterior margin of mouth does not extend past eye . . . . smallmouth bass (*Micropterus dolomieu*)

26. (a) Dorsal surfaces are mottled . . . . white perch (*Morone americana*)

    (b) Dorsal surfaces are striped . . . . go to 27

27. (a) Anterior margin of anal fin is squared . . . . striped bass (*Morone saxatilis*)

    (b) Anterior margin of anal fin is rounded . . . . white bass (*Morone chrysop*)

Key Terminology

barbel—a fleshy sensory structure on the snout of a fish; a "whisker"

dorsal—back, top

heterocercal—a tail with unequally sized lobes (The dorsal lobe is usually larger, and the vertebral column may extend into it.)

homocercal—a tail with equal-sized lobes

inferior—beneath the snout

posterior—rear

scute—a bony plate

soft ray—a soft, flexible bone that supports a fin

spiny ray—a stiff, sharp bone that supports a fin

subterminal—slightly below the snout

terminal—at the end of the snout

vermiculated—wormlike

ventral—underside, bottom

Additional fish external anatomy terms can be found on pages 402–3 in your textbook.

11B LAB

All Myxed Up

A Case Study in Classification

How does taxonomy enhance our understanding of living things? People classify things all the time, and usually for a very good reason: categorized things are easier to handle. Think of a mechanic doing a tune-up on the family car. If the mechanic needs a particular wrench to loosen the oil pan plug, the job could be done more efficiently if all the wrenches are sorted by size in one drawer of the tool chest rather than tossed together willy-nilly.

Taxonomy is the science of classifying living things, and taxonomists classify living things for much the same reason as mechanics classify tools. Living things are easier to study and discuss if they have been grouped into categories on the basis of shared characteristics. Exactly how taxonomy has done this, however, has changed over the years, and the reason why it has changed has significance for scientists—and biology students—who believe in the truth of God's Word.

In this lab activity you will read portions of Carl Linnaeus's *Systema Naturae* along with the abstract from a modern scientific paper. An *abstract* is a paragraph that sums up a journal article or research paper, including the research question and the answer that the researcher found. The abstract that you will read is from a paper on the classification of myxozoans.

Myxozoans are microscopic marine and freshwater parasites whose main primary hosts are fish. The first myxozoan was identified in 1825, but myxozoans were not taxonomically classified until 1881. Since then, the taxonomy of myxozoans has undergone at least four major revisions, along with many minor revisions, and yet there is still disagreement on how they should be classified. You may not understand all the terminology in the following passages, but don't be too anxious about the specifics—we're mainly interested in the central ideas. Your goal is to consider what classification is good for and what it's *not* good for.

Why do scientists classify things?

Equipment
none

QUESTIONS
- What is taxonomy?
- How has the science of taxonomy changed?

PROCEDURE

Carl Linnaeus and Systema Naturae

Carl Linnaeus is considered the Father of Modern Taxonomy. In 1735 he published *Systema Naturae*, the first thorough attempt to classify living things. In it Linnaeus introduced the system of *binomial nomenclature*, which identifies the genus and species of the organism and is still used today for naming all life-forms. What motivated Linnaeus to write his best-known work? Let's read through some of his own words from the introduction to the first edition of *Systema Naturae*. (*Note:* The paragraphs were numbered in the original. Portions of the original are omitted here, so some numbers in the sequence are missing.)

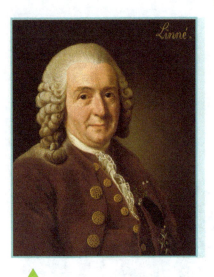

Carl Linnaeus

Observations on the Three Kingdoms of Nature

1. If we observe God's works, it becomes more than sufficiently evident to everybody that each living being is propagated from an egg and that every egg produces an offspring closely resembling the parent. Hence no new species are produced nowadays.

4. As there are no new species; as like always gives birth to like; as one in each species was at the beginning of the progeny, it is necessary to attribute the progenitorial unity to some Omnipotent and Omniscient Being, namely God, whose work is called Creation. This is confirmed by the mechanism, the laws, principles, constitutions and sensations in every living individual.

7. On our earth, only two of the three mentioned above [*Note:* Section 6 listed celestial bodies, elements, and natural bodies] are obvious; i.e., the elements constituting it; and the natural bodies constructed out of the elements, though in a way inexplicable except by creation and by the laws of procreation.

8. Natural objects belong more to the field of the senses than all the others and are obvious to our senses anywhere. Thus, I wonder why the Creator put man, who is thus provided with senses and intellect, on the earth globe, where nothing met his senses but natural objects, constructed by means of such an admirable and amazing mechanism. Surely for no other reason than that the observer of the wonderful work might admire and praise its Maker.

10. The first step in wisdom is to know the things themselves; this notion consists in having a true idea of the objects; objects are distinguished and known by classifying them methodically and giving them appropriate names. Therefore, classification and name-giving will be the foundation of our science.

12. He may call himself a naturalist (a natural historian), who well distinguishes the parts of natural bodies by sight and describes and names all these rightly in agreement with the threefold division. Such a man is a lithologist, a phytologist, or a zoologist.

17. I have shown here a general survey of the system of the natural bodies so that the curious reader with the help of this, as it were, geographical table knows where to direct his journey in these vast kingdoms, for to add more descriptions, space, time, and opportunity lacked.

CAROLUS LINNAEUS
Doctor of Medicine
Given at Leyden, July 23, 1735[1]

Answer the following questions on the basis of your reading of the introduction to *Systema Naturae*. Cite the sections that support your answer for each question.

1. What evidence do you find in Linnaeus's introduction that suggests that he had a biblical worldview?

_____

_____

_____

_____

_____

---

[1] Carolus Linnaeus, *Systema Naturae* (1735; Facsimile of the First Edition), trans., M. S. J. Engel-Ledeboer & H. Engel (Nieuwkoop, Netherlands: B. De Graaf, 1964), 18–19

2. Did Linnaeus believe that organisms could change over time? Explain.

_____

_____

_____

3. What justification did Linnaeus give for studying classification?

_____

_____

_____

_____

_____

4. What criterion did Linnaeus use to classify "natural bodies" into his system?

_____

_____

_____

5. What reason does Linnaeus give for publishing *Systema Naturae*?

_____

_____

_____

Although not given by Linnaeus as a reason for publishing his book, one of the long-term effects of the widespread acceptance of *Systema Naturae* was the universal adoption of binomial nomenclature, in which each species of organism described by science is given a unique scientific name. The mountain lion (also known as the painter, panther, cougar, catamount, ghost cat, fire cat, puma, ko-icto, katalgar, coowachobee, or klandagi), for example, is *Puma concolor*, regardless of whether a scientist studying one is from Portland or Puerto Vallarta.

6. How does the assigning of scientific names make it easier for scientists to study and communicate about organisms?

_____

_____

_____

The Myxozoa and Modern Classification

Carl Linnaeus died eighty-one years before Charles Darwin published *On the Origin of Species*. In Linnaeus's day most scientists believed in a literal reading of Genesis. They also held to the Aristotelian idea of the fixity of species, the idea that organisms do not change over time (see page 200 in your textbook). Since then, however, most scientists have embraced an evolutionary worldview that rejects both of these ideas. The effects of this change in worldview on the science of taxonomy have been striking. To see some of these effects, take a look at the following abstract written by some modern taxonomists.

FROM "GENOMIC INSIGHTS INTO THE EVOLUTIONARY ORIGIN OF MYXOZOA WITHIN CNIDARIA"

"The Myxozoa comprise over 2,000 species of microscopic obligate parasites that use both invertebrate and vertebrate hosts as part of their life cycle. Although the evolutionary origin of myxozoans has been elusive, a close relationship with cnidarians, a group that includes corals, sea anemones, jellyfish, and hydroids, is supported by some phylogenetic studies and the observation that the distinctive myxozoan structure, the polar capsule, is remarkably similar to the stinging structures (nematocysts) in cnidarians. To gain insight into the extreme evolutionary transition from a free-living cnidarian to a microscopic endoparasite, we analyzed genomic and transcriptomic assemblies from two distantly related myxozoan species, *Kudoa iwatai* and *Myxobolus cerebralis*, and compared these to the transcriptome and genome of the less reduced cnidarian parasite, *Polypodium hydriforme*. A phylogenetic analysis, using for the first time to our knowledge, a taxonomic sampling that represents the breadth of myxozoan diversity, including four newly generated myxozoan assemblies, confirms that myxozoans are cnidarians and are a sister taxon to *P. hydriforme*. Estimations of genome size reveal that myxozoans have one of the smallest reported animal genomes. Gene enrichment analyses show depletion of expressed genes in categories related to development, cell differentiation, and cell–cell communication. In addition, a search for candidate genes indicates that myxozoans lack key elements of signaling pathways and transcriptional factors important for multicellular development. Our results suggest that the degeneration of the myxozoan body plan from a free-living cnidarian to a microscopic parasitic cnidarian was accompanied by extreme reduction in genome size and gene content."[2]

Answer the following questions on the basis of your reading of the abstract. Cite examples from the abstract to support your answers.

7. Do the authors of this abstract have a biblical worldview or an evolutionary worldview? Explain.

_____

_____

_____

_____

8. In contrast to the morphology-based taxonomies of previous efforts to classify the myxozoans, what does the abstract state that the most recent revisions are based on?

_____

_____

_____

_____

---

[2] E. Sally Chang, et. al., "Genomic Insights into the Evolutionary Origin of Myxozoa within Cnidaria," Dec. 1 2015 (Washington, DC: National Academy of Sciences).

ANALYSIS

9. Why did Linnaeus classify organisms on the basis of appearance?

_____

_____

_____

_____

10. Why do evolutionists classify organisms on the basis of phylogeny?

_____

_____

_____

_____

11. How do these two different methods affect how often classification changes?

_____

_____

_____

_____

_____

_____

12. Which classification method is more practical? Explain.

_____

_____

_____

_____

_____

_____

_____

_____

13. How would Linnaeus have classified myxozoans and cnidarians? Explain.

_____

_____

_____

14. How do modern taxonomists classify myxozoans and cnidarians?
    (*Hint:* Refer to the abstract that you just read.)

    _____

    _____

    _____

15. Can modern phylogeny-based taxonomic systems ever produce a satisfactory
    phylogenetic tree that is based on DNA analysis? Explain.

    _____

    _____

    _____

    _____

    _____

    _____

GOING FURTHER

16. Think back to what you learned about evolution in Chapter 10. Using what
    you know about evolutionary theory and what you read in the abstract
    above, why do you think myxozoans are actually a bad example of supposed
    evolution in action?

    _____

    _____

    _____

    _____

    _____

    _____

    _____

    _____

    _____

    _____

12A LAB

Squeaky Clean

Bacteria Growth and Handwashing

"Make sure you wash your hands with soap and warm water." If you had a dollar for every time you heard these words, you'd probably be rich! But do you know why you need to wash your hands this way?

People haven't always had medical reasons to justify handwashing. Though some cultures, such as the Jews (Matt. 15:1–2), practiced ritual handwashing, it didn't become widespread for medical purposes until the 1800s.

Handwashing saves people's lives. In 1845 Hungarian physician Ignaz Semmelweis pioneered the practice of doctors washing their hands before working with patients after he observed that handwashing greatly reduced the occurrence of dangerous infections in new mothers. Although many doctors rejected his practices, they eventually became standard, and Semmelweis was vindicated years after his death by the formation of the germ theory of disease.

Since the bacteria on your hands are too small to see, in this lab activity you will be growing a colony that will be large enough for you to see without a microscope. To do this, you will use a *petri dish*, a large flat dish with a lid (see Appendix B), containing a gel-like substance called *nutrient agar*. Nutrient agar is a mixture of beef extract, a protein called *peptone*, water, and a gel derived from algae. Nutrient agar provides both the structure and the nutrients that bacteria need to grow into a colony.

How do different methods of handwashing affect bacteria count?

Equipment

microscope

prepared sterile petri dish with cover

ruler

marker

sterile cotton swabs (4)

hand soap

hand sanitizer

plastic bag, large

laboratory apron

goggles

QUESTIONS

- How dangerous is it when I don't wash my hands?

- What does water do in handwashing?

- What does soap do in handwashing?

QUARTERS OF YOUR PETRI DISH

1. Nothing applied to agar

2. Dirty hand applied to agar

3. Hand-washed only with water applied to agar

4. Hand-washed with both soap and water applied to agar

▲

Follow the pattern above when swabbing your petri dish, but keep your swabbing in the appropriate area.

PROCEDURE

Starting Bacterial Cultures

Your teacher will provide you with a prepared, sterilized petri dish containing solidified nutrient agar.

A Turn the petri dish over. Divide the dish into quarters by drawing on the bottom of the petri dish with the ruler and marker. Label the quarters 1, 2, 3, and 4.

B Shake hands with two other people in your group. You will be testing the bacteria count on each person's hands in a different way.

C Swab one of your group member's hands with a sterile cotton swab. Without putting the swab on the benchtop, open the dish just enough to be able to insert the swab. Now swab Area 2 of your petri dish, using a zigzag squiggle. (See margin, below left.) Turn the dish 90° and do it again, still in Area 2. Close your dish as soon as possible.

D Have one person from your group wash his hands with only warm water for 15 seconds. He should dry his hands with a paper towel. Swab his hand with another sterile cotton swab and apply it to Area 3 using the same process as you did in Step C.

E Have one person from your group wash his hands with warm water and hand soap for fifteen seconds. He should dry his hands with a fresh paper towel. Swab his hand with another sterile cotton swab and apply it to Area 4 using the same process as you did in Step C.

1. Why do all of the people whose hands will be tested need to shake hands?

   _____

   _____

2. What does your answer to Question 1 tell you about the cultural practice of shaking hands with people?

   _____

   _____

3. Why should you avoid setting your swab on the benchtop?

   _____

   _____

4. Why should you open the petri dish only partway and close it as quickly as possible?

   _____

   _____

F Now put some hand sanitizer on a fresh, sterilized cotton swab and swab along the two lines separating the quarters.

G Label a plastic bag with your group's name. Place your petri dish in the plastic bag and seal it. You will not remove the petri dish from the bag for the rest of this lab activity. Your teacher will store this for a few days until the bacterial colonies grow large enough for you to observe.

H Everyone in your group must wash their hands.

5. What is the function of Area 1 if you haven't added anything to the agar?

_____

_____

6. If a bacterial colony grows in Area 1, what would you conclude?

_____

_____

7. What do you expect to see growing in each quadrant of your petri dish? Explain your reasoning.

_____

_____

8. Why did you swab the lines separating the quadrants with hand sanitizer?

_____

_____

Observing Sample Bacteria

I Observe the samples of bacteria provided by your teacher. These are some bacteria that you may see in your petri dish.

9. Which bacteria did you observe?

_____

_____

10. Describe what you observed and classify the bacteria shape.

_____

_____

_____

When scientists observe bacteria, they need to stain them in a way that makes them easier to observe and helps to identify whether the bacterium's cell walls contain peptidoglycan. If a bacterium's cell wall contains the protein, it looks purple when stained. This kind of bacterium is referred to as *Gram-positive*. If a bacterium's cell wall doesn't contain peptidoglycan, it looks pink when stained, and it is referred to as *Gram-negative*.

11. How does the information about Gram staining relate to what you observed?

_____

_____

_____

Gram-positive bacteria

Observing Your Bacteria

After several days of incubation, your teacher will return your petri dish. ***Do not remove the dish from the bag.*** Observe whichever side gives you a better view of the bacterial colonies.

J Count the number of colonies, or clumps of bacteria, in each quadrant and record your data in Table 1.

TABLE 1

Quadrant	Number of Colonies
1	
2	
3	
4	

12. What colors and textures did you observe in your bacteria colony? What do these different colors represent?

_____

_____

_____

_____

13. Using the table on the next page, describe the form, elevation, and margins of the bacteria you grew.

_____

_____

_____

_____

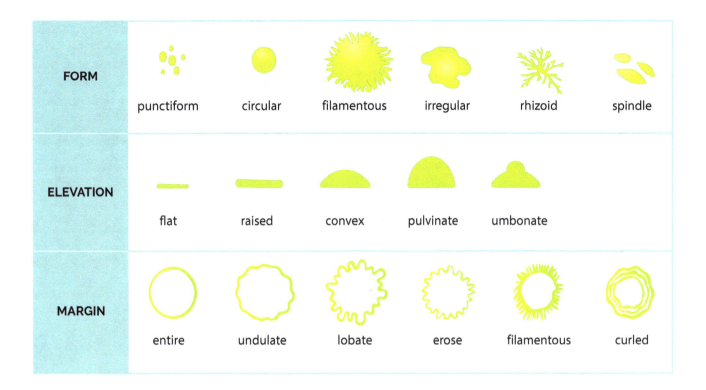

FORM	punctiform	circular	filamentous	irregular	rhizoid	spindle
ELEVATION	flat	raised	convex	pulvinate	umbonate	
MARGIN	entire	undulate	lobate	erose	filamentous	curled

ANALYSIS

14. If possible, identify any of your bacterial colonies on the basis of your observations. You may use the internet to help you.

_____

_____

_____

K Use a ruler to measure the larger bacterial colonies.

15. Which bacterial culture appeared to grow the fastest? How can you tell?

_____

_____

_____

16. How did your hypothesis in Question 7 compare with what you observed?

_____

_____

_____

17. Did you find any bacterial colonies in Area 1? How does this affect the results that you observed in the other areas of your petri dish?

_____

_____

_____

_____

L Return your petri dish to your teacher for disposal and then everyone in the group must wash their hands.

CONCLUSION

18. On the basis of what you have observed, why do you think it is important to wash hands properly?

_____

_____

_____

19. What does washing hands with just warm water do?

_____

_____

_____

20. Sometimes soap in dispensers in restrooms becomes contaminated with bacteria. How could you tell whether the soap you were using in this experiment was contaminated?

_____

_____

_____

21. Suggest what happens to bacteria during handwashing. How is this different from using hand sanitizer?

_____

_____

_____

_____

_____

_____

_____

Handwashing saves lives. When Ignaz Semmelweis began practicing handwashing, he watched more of his patients survive to care for the children they had just birthed. The graph below shows the percent of patients who died from puerperal fever at the First Clinic of Vienna Maternity Institution for the years 1841–49.

22. When did Semmelweis notice the biggest decrease in mortality rates in his maternity patients?

_____

23. Explain what appears to have caused this reduction in mortality.

_____

_____

24. Why is it worthwhile to put scientific research into the practice of handwashing?

_____

_____

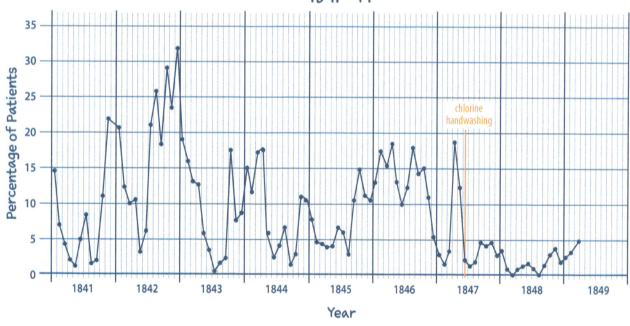

Puerperal Fever Monthly Mortality Rates 1841– 49

GOING FURTHER

You can take this experiment further by testing how other factors affect bacterial counts on our hands. Create your own experiment from the following ideas.

> » Test other methods of handwashing, including using dirt or ash as is done in some poorer countries.
>
> » Test different kinds of soaps including antibacterial soaps.
>
> » Test how the amount of time lathering affects bacterial count.
>
> » Test whether water temperature affects the results of handwashing.

M You collect lots of bacteria on your hands by touching other people! Physical contact with people is one source of bacteria. For another experiment, try swabbing different places in your home or classroom and culture these samples to find the places in your environment where bacteria make a home.

25. What will you test?

_____

_____

26. Write out the plan for your experiment.

_____

_____

_____

_____

_____

_____

N Report the results from your experiment in a separate report or presentation.

12B LAB

One Slick Solution

Oil-Eating Bacteria

Deep, deep down in the murky depths of the Gulf of Mexico there was a big problem. The Deepwater Horizon oil rig operated by British Petroleum had exploded and sunk on April 20, 2010, killing eleven men and causing a gusher on the ocean floor that leaked oil for eighty-seven days. It was one of the largest oil spills in history. Oil spills are dangerous because oil pollutes the environment and is toxic to living things, especially sea birds, turtles, and dolphins. It took months for a huge team including scientists, volunteers, and robots to cap the leak and clean up miles of ocean and coastline.

Volunteers put out booms to contain the floating oil and cleaned beaches and wildlife. Aircraft pilots skimmed oil off the surface of the ocean. One of their smallest allies may surprise you—bacteria! Scientists injected the genetically engineered bacteria *Alcanivorax borkumensis* near the leak on the sea floor. The bacteria actually ate some of the oil before it floated to the surface.

Eliminating the oil before it rose to the surface was important because it protected wildlife and prevented currents from carrying the oil to beaches in Louisiana, Alabama, and Florida. Using nature to clean up the environment in this way is called *bioremediation*. It's one of the most efficient ways to restore a contaminated environment.

Oil from natural sources regularly seeps into the ocean environment—though not 210 million gallons! Oil is a long molecule made from mostly carbons and hydrogens. Many microbes use oxygen to naturally break down these long molecules into byproducts like carbon dioxide and water that are harmless to living things. It's a process that is similar to cellular respiration, but it uses hydrocarbons instead of glucose.

?

How can we use our knowledge of bacteria to clean up oil spills?

The US Coast Guard and British Petroleum also used controlled burning of oil on the water's surface to clean up the Deepwater Horizon spill.

Pseudomonas, an oleophilic bacteria

Bacteria that process oil are known as *oleophilic bacteria*. In this lab activity you will be working with a mix of different types of oil-eating bacteria to see how they can break down oil into smaller molecules to reduce the amount of floating oil. But to survive, the oil-eating bacteria will need air and some nutrients, such as potassium and nitrogen. During remediation efforts, scientists sometimes add these nutrients to polluted environments in the form of a fertilizer, which works well when cleaning contaminated soil. But this isn't very effective in the open ocean.

Oleophilic bacteria can live in a pH range of 5.5–10 and at a temperature ranging from −2 °C to 60 °C. You'll see how these other factors help bacteria grow so that they can consume more oil.

1. Write a chemical equation, showing the reactants and products, in which oleophilic bacteria metabolize oil. Though petroleum is a mix of hydrocarbons, use octane (C_8H_{18}) in your chemical formula. Bacteria use oxygen to break down octane into water and carbon dioxide.

_____

Equipment

laboratory balance (accurate to 0.01 g)

graduated cylinders, 50 mL (4)

graduated cylinder, 10 mL

glass stirring rods (3)

wax pencil

weighing paper

cotton balls

disposable plastic pipettes

oil-eating bacteria suspension, 80 mL

10-10-10 fertilizer

acetic acid (CH_3COOH), 1 M, 5 mL

universal indicator paper or pH meter

milk of magnesia, 8 mL

cooking oil, 60 mL

laboratory apron

nitrile gloves

goggles

QUESTIONS

- What do bacteria need to grow?

- How can we nurture oil-eating bacteria to clean up an oil spill?

PROCEDURE

Setting Up

A Using a wax pencil, label four graduated cylinders from 1 to 4. Add 20 mL of the bacterial suspension to each graduated cylinder.

B Weigh out 0.25 g of fertilizer on three different pieces of weighing paper. Add this fertilizer to Cylinders 2, 3, and 4. Use a separate glass stirring rod to stir the solutions in each graduated cylinder until the fertilizer is dissolved.

2. Predict whether the fertilizer will encourage or inhibit bacterial growth. Explain.

_____

_____

_____

C Use the 10 mL graduated cylinder to add 5 mL of acetic acid to Cylinder 3 and stir the contents. Determine the pH of the solution using universal indicator paper or a pH meter.

NUMBERING YOUR GRADUATED CYLINDERS

1. bacterial suspension, oil

2. bacterial suspension, oil, fertilizer

3. bacterial suspension, oil, fertilizer, acetic acid

4. bacterial suspension, oil, fertilizer, milk of magnesia

3. What is the pH that you measured? On the basis of your measurement, predict whether the acetic acid will encourage or inhibit bacterial growth. Explain.

_____

_____

_____

D Add 8 mL of milk of magnesia to Graduated Cylinder 4 and stir the contents. Determine the pH of the solution using universal indicator paper or a pH meter.

4. What is the pH that you measured? On the basis of your measurement, predict whether the milk of magnesia will encourage or inhibit bacterial growth. Explain.

_____

_____

_____

E Add 15 mL of cooking oil to all four graduated cylinders.

5. What is the purpose of Graduated Cylinder 1?

_____

_____

6. Why does the oil float to the top when you add it to each graduated cylinder?

_____

_____

7. How would this property of oil affect how it is cleaned up in a spill?

_____

_____

F Measure the volume of the oil layer in each graduated cylinder by subtracting the volume at the bottom of the oil layer from the volume at the top of the oil layer. (Refer to Appendix D if you need help on how to read a graduated cylinder.) Record these values to at least 0.1 mL precision in the _Day 0_ column of Table 1.

G Put a cotton ball in the top of each graduated cylinder to prevent contaminants from entering it.

H Place the cylinder in the location specified by your teacher. Wash your hands before leaving the laboratory.

8. Predict which of the graduated cylinders will show the most bacterial growth. Explain your prediction.

_____

_____

_____

I After Day 1, read the volumes at the top and bottom of each oil layer to determine the volume of the oil to at least 0.1 mL precision. Record your data in Table 1.

J After measuring the volume of each oil layer, aerate each cylinder. Use a separate pipette for each cylinder. Insert the pipette into the lower liquid layer, then squeeze the pipette bulb, forcing air into the liquid. Do not release the bulb until it has been removed from the cylinder. Make sure that you aerate all graduated cylinders to the same degree.

K Return the cylinder to its safe location. Wash your hands before leaving the laboratory.

L Repeat Steps I–K for the next three days.

9. What would happen if you didn't bubble air through the graduated cylinders?

_____

_____

10. Why should you use a separate pipette for each graduated cylinder instead of using only one pipette for all of the graduated cylinders?

_____

_____

11. Did you observe any changes in the appearance of the liquid in the graduated cylinders? Explain.

_____

_____

M Dispose of your graduated cylinders according to your teacher's instructions.

ANALYSIS

N Use Graphing Area A to graph your data from Table 1. Plot the time in days on the *x*-axis and the volume of the oil layers in milliliters on the *y*-axis. Use a different color for each graduated cylinder.

12. Was your prediction in Question 8 accurate? Explain.

_____

_____

_____

_____

13. Write down any surprising observations and try to explain them.

_____

_____

_____

_____

14. If you had not equally aerated your graduated cylinders, how would your results have been affected?

_____

_____

_____

_____

CONCLUSION

15. On the basis of what you have observed, what do you think is necessary for bacteria to be helpful in cleaning up oil spills?

_____

_____

_____

Of the oil spilled in the Deepwater Horizon oil spill, 17% was captured from the well and 8% was removed through skimming and burning. Workers also successfully used chemicals to disperse the oil. One of the reasons they did this was to make it easier for bacteria to metabolize the oil. It takes a long time for bacteria to consume this much oil.

16. Suggest one limitation or negative effect of bioremediation for an oil spill using oleophilic bacteria.

_____

_____

_____

17. In what ways did your graduated cylinders model a real oil spill? What variables were left out of your model?

_____

_____

_____

_____

_____

People have compared the 2010 Deepwater Horizon oil spill to the *Exxon Valdez* oil spill in 1989. Though the Deepwater Horizon event was a larger spill, it contained a lighter oil that had shorter hydrocarbons. The *Exxon Valdez* oil spill happened on the ocean surface and closer to the Alaskan shoreline, resulting in more injured animals. Also, the Gulf of Mexico has more natural seeps, so there were naturally more oleophilic bacteria in the water. Finally, the Deepwater Horizon spill happened in an area of the world where the temperature is much more nurturing to bacteria than where the *Exxon Valdez* spill occurred.

But one of the most significant differences in these two incidents was the massive response of employees and volunteers to clean up the Deepwater Horizon spill, involving 47,000 people. This oil spill was in a much more heavily populated area that was easier to access, resulting in an overwhelming response from the local community.

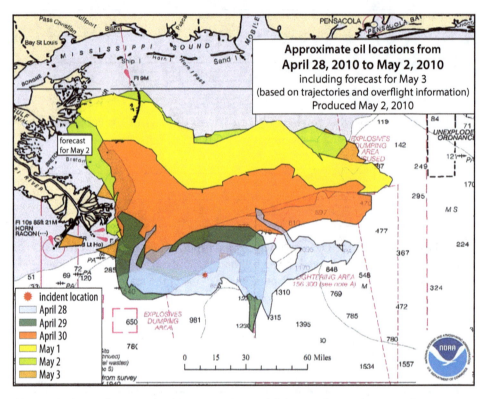

Meteorologists and ocean scientists team up to model the way that environmental factors affect an oil spill in order to predict how the oil will spread.

18. How is the conservation work involved in cleaning up an oil spill glorifying to God?

_____

_____

_____

_____

GOING FURTHER

19. You varied the pH in the graduated cylinders to see how it affected the growth of oleophilic bacteria. What other things could you have varied?

_____

_____

_____

20. Choose a variable from your answer to Question 19 and write out a plan to test it. If you'd like to actually carry out this experiment, collect data and share it with others.

_____

_____

_____

_____

_____

TABLE 1

GRADUATED CYLINDER	VOLUME OF OIL LEVEL (mL)				
	Day 0	Day 1	Day 2	Day 3	Day 4
1					
2					
3					
4					

GRAPHING AREA A

13A LAB

Wee, Watery World

Exploring the Microscopic World of Protozoans

Hundreds of thousands of idealistic young men fell during the American Civil War. The majority of those who perished were not killed by bullets or shells, but by disease. One of the most common diseases they suffered from was *amoebic dysentery* caused by the parasitic protozoan *Entamoeba histolytica*. The spores of *E. histolytica* survive in contaminated soil and water. Unsuspecting soldiers ingested the spores when they drank from contaminated streams.

Many other protozoans are free-living (nonparasitic) and can be found in water and soil around the globe. They are an important part of aquatic food webs. Protozoans exhibit an astonishing variety of forms and functions. In this lab activity you'll examine some of these amazing—and sometimes deadly—microscopic organisms.

What are the distinguishing features and characteristics of protozoans?

Equipment

microscope	cover slips (5)	toothpick
reference books or keys for protozoans	living cultures of amoebas, euglenas, and paramecia	carmine powder or yeast
pipettes (5)	cotton fibers	pond water
microscope slides (5)	glycerin or methyl cellulose	

OBSERVING PROTOZOANS

Amoeba

A Obtain a sample from the amoeba culture and prepare a wet mount of it (see Appendix D). Amoebas usually stay close to the bottom of the culture or crawl on some object, but they can be drawn into a pipette by using the following technique.

» Squeeze the air out of a pipette bulb.

» Place the pipette directly above the place where the amoeba should be.

» Release the bulb to suck up one drop.

» Quickly put the entire drop on the slide. (Wait a minute before placing the cover slip on top of the amoeba culture on your slide to allow the amoeba to attach to the slide and begin to move.)

QUESTIONS

- What are protozoans?
- What are some of the different kinds of protozoans?
- How do protozoans move and get food?
- What kinds of protozoans can be found in pond water?

NOTES ON HANDLING LIVE CULTURES

1. Live cultures often contain a few organisms other than the targeted species.

2. Live cultures sometimes go "sour," meaning that the organisms die. If this happens, observe a prepared slide of the desired organism.

3. Keep the cultures covered when not in use.

4. Use a separate pipette for each culture to prevent cross contamination.

5. Make sure that your slides and cover slips are clean and free of any soap residue.

B Scan the entire slide on low power. Scan first from right to left; then move the slide a little lower and scan left to right. Repeat until you have scanned the entire slide. Anything that moves faster than a snail's pace is *not* an amoeba.

C Be sure that the culture medium does not evaporate. Using the pipettes from the amoeba culture, add more medium to the edge of the cover slip as necessary.

D Observe the movements of the amoeba for a while on both low and high power.

E If you see an amoeba engulfing food or dividing, inform your teacher so that you can share it with the class.

1. Describe the locomotion (movement) of the amoeba that you observed.

   _____

   _____

2. What is the name for this type of locomotion?

   _____

3. Is more than one pseudopod present at one time?

   _____

4. Besides movement, for what other purpose does an amoeba normally form pseudopods?

   _____

F Draw an amoeba in Drawing Area A. Label as many structures as you can.

Euglena

G Prepare a wet mount from the euglena culture. Euglenas will be found throughout the culture medium, so no special capture technique is necessary.

H Scan the slide. Euglenas can move quickly but usually will not leave the microscope field very rapidly. Occasionally it will be necessary to use cover slip pressure or a thicker medium to slow their movement.

I Observe the euglenas on both low and high power.

5. Describe how euglenas move.

   _____

   _____

6. Were you able to observe a second kind of movement? If so, describe it.

   _____

   _____

7. Scientists have disagreed over the years about how to classify euglenas, sometimes grouping them with protozoans, and at other times classifying them as algae. What characteristics of euglenas are plantlike? Which are animal-like?

_____

_____

_____

_____

J Draw a euglena in Drawing Area B. Label as many structures as you can.

Paramecium

K Prepare a wet mount from the paramecium culture. Paramecia may be found throughout the culture.

L Scan the entire slide. Paramecia move rapidly and will need to be chased across the slide. To slow down or stop the paramecia, your teacher may have you take one of the following actions.

» Have your lab partner use the edge of a paper towel to wick away a small amount of culture medium from beneath the cover slip while you keep the paramecium in your field of view. As the culture is drawn off, the cover slip will exert more pressure on the paramecium. Be careful to not allow the medium to evaporate completely.

» Before placing the cover slip, place a small quantity of cotton fibers on the slide. These will block the paramecia's paths and localize their activity.

» Use a special medium prepared with glycerin or methyl cellulose. Because these media are thicker than the culturing solution, protozoans move more slowly in them.

Be careful not to confuse other protozoans in the culture with paramecia. A paramecium looks like the slipper-shaped illustration seen on page 277 in your textbook and is easy to recognize.

M Observe the paramecium on both low and high power and try to locate all the structures pointed out in your textbook.

N After observing your paramecium for a while, carefully remove the cover slip from your slide and add a few cotton fibers to the medium if you have not done so already; do this only if you are not using the thickened medium.

O Observe the reactions of the paramecia as they encounter the cotton fibers.

8. Describe the movement of the paramecium. Be sure to include the names of the cellular structures involved in its movement.

_____

_____

9. How does the paramecium react when it encounters a cotton fiber?

_____

_____

_____

10. How does amoeboid movement compare to paramecium movement?

_____

_____

_____

11. Look carefully at a resting or confined paramecium and observe the operation of its contractile vacuole. Describe what you see.

_____

_____

_____

12. Could a paramecium survive if its contractile vacuole malfunctioned? Explain.

_____

_____

_____

_____

_____

P Using a toothpick, place a few grains of carmine powder or yeast on your slide. Observe the path of the powder or yeast as it enters the paramecium and moves within it.

13. What is the function of the carmine powder?

_____

_____

14. How do paramecia obtain their food?

_____

_____

15. How is the paramecium's food digested?

_____

_____

Q Draw a paramecium in Drawing Area C. Label as many structures as you can.

OBSERVING POND WATER

A drop of pond water is often an astonishing hive of activity. The typical pond is home not only to amoebas, euglenas, and paramecia, but a host of other protists as well. These organisms are an important part of a pond's food web.

R Make two wet mounts from the pond water culture provided by your teacher: one from the material near the bottom of the culture and the other from material near the top.

16. Why is it advisable to take samples from both areas?

_____

_____

_____

S Use Drawing Area D to draw some of the organisms that you observe in your mounts. Reference the books and keys provided by your teacher to identify each organism, if possible, including the taxonomic groups that each one belongs to (e.g., Protozoa). Label any structures that you are able to identify.

GOING FURTHER

Entamoeba histolytica wasn't identified until 1873, well after the Civil War ended. Later, scientists and doctors noted that most people infected with *E. histolytica* didn't actually get sick. The search for the answer to why that was so is one of the many obscure but interesting stories of science.

T Research information about Émile Brumpt; then answer the following questions.

17. What did Émile Brumpt suggest was the reason why most people did not get sick when infected with *E. histolytica*?

_____

_____

_____

_____

18. Why was Brumpt's proposal not initially accepted by the scientific and medical communities?

_____

_____

_____

_____

19. How was the question of the existence of a nonpathogenic *Entamoeba* finally resolved?

_____

_____

_____

_____

20. What lesson can you draw about the nature of science from the story of *Entamoeba*?

_____

_____

_____

_____

21. What practical applications might our better understanding of *Entamoeba* have?

_____

_____

_____

_____

22. Using what you learned about the study of *E. histolytica*, how do you think the Creation Mandate is further advanced by the continued study of organisms that we already seem to know much about?

_____

_____

_____

_____

_____

_____

_____

_____

DRAWING AREA A

DRAWING AREA B

DRAWING AREA C

DRAWING AREA D

13B LAB

Fun with Fungi

Observing Fungi

A group of hikers are returning to their vehicle at dusk. Suddenly they see a faint, eerie glow just a few feet in front of them near the ground by the trail. Shining a flashlight toward the glow reveals only a rotting stump covered with shelf fungi. When the light is turned off, the glow vanishes.

These hikers aren't imagining things. The shelf fungi growing on that stump are *Panellus stipticus*, a species of glowing fungi. The glow is very faint, so when the hikers shined their flashlight on the stump, it reflected enough light to their eyes that they could no longer see the faint glow until they had once again readjusted to the dim light.

Fungi may seem irrelevant since we seldom see them unless they have large fruiting structures, but they are very important to life. They are responsible for most of the decomposition of dead organisms—imagine the mess we would have without decomposers! Fungi also include edible mushrooms, yeasts used for baking bread, and molds used for some cheeses, such as blue cheese. Lunch wouldn't be nearly as much fun without fungi!

What distinguishes a fungus from other organisms and from other fungi?

Equipment

- dissecting microscope or hand lens
- microscope
- living cultures of *Rhizopus stolonifer* and *Penicillium notatum*
- preserved slides of *Rhizopus stolonifer*, w.m.; cup fungus apothecium; and *Coprinus*, c.s.
- preserved specimens of puffballs, shelf fungi, and mushrooms
- laboratory apron
- goggles

QUESTIONS

- What are the characteristics of fungi?
- How do the phyla of fungi differ?
- How are fungi classified?

PROCEDURE

Phylum Zygomycota

The phylum Zygomycota derives its name from the thick-walled sexual structures called *zygosporangia* that are produced when specialized hyphae unite. These fungi are usually sessile, produce spores, and resemble algae in structure. In addition to the sexually produced zygosporangia, zygomycota bear asexually produced spores in sporangia. Examples of this phylum are common and abundant in the soil and air and include molds and yeast. One, *Rhizopus stolonifer*, is a black mold that often grows on bread.

A Using either a dissecting microscope or a hand lens, observe cultures of *R. stolonifer*.

1. Describe what you see, including any fungal structures that you can identify.

_____

_____

2. *R. stolonifer* produces hyphae that are clear or white. What causes its dark appearance?

_____

B Observe a preserved slide of *R. stolonifer* on high dry power.

C Observe a preserved slide of *R. stolonifer* with zygotes.

D Use Drawing Area A to draw a zygosporangium with two parent hyphae.

Phylum Ascomycota

Fungi in phylum Ascomycota often appear very similar to fungi in phylum Zygomycota but differ in their spore-forming structures. They are named for their microscopic sac-like asci (s. ascus) that hold the sexually produced ascospores.

E Using either a dissecting microscope or a hand lens, observe the laboratory culture of *Penicillium notatum*.

3. Describe what you see, including any fungal structures that you can identify.

_____

_____

F Observe a preserved slide of a cup fungus apothecium.

4. Describe the asci that line the apothecium.

_____

_____

G Use Drawing Area B to draw one of the asci that line the apothecium.

Phylum Basidiomycota

As you learned in your textbook, phylum Basidiomycota includes mushrooms, puffballs, and shelf fungi. The phylum's name comes from the sexually produced basidiospores borne on club-shaped cells called *basidia* (s. basidium). On the familiar mushroom, thousands of basidia are attached to the gills under the mushroom's cap, and each basidium has four basidiospores.

H Using either a dissecting microscope or a hand lens, observe preserved and dried specimens of puffballs.

5. Where are the basidia and basidiospores located on puffballs?

_____

6. How are the basidiospores released?

_____

_____

I Using either a dissecting microscope or a hand lens, observe preserved specimens of shelf fungi.

7. Where are the basidia and basidiospores located on shelf fungi?

_____

8. How are spores released from shelf fungi?

_____

J Using a dissecting microscope or a hand lens, observe preserved specimens or fresh samples of mushrooms.

» Note the cap, gills, stipe, and hyphae of the various species.

» Note the different colors, sizes, and textures of the various species.

K Observe the preserved slide of *Coprinus*. Be sure to observe the cap.

» Find the gills and locate the basidia and basidiospores.

L Use Drawing Area C to draw a section of a gill with basidia and basidiospores.

9. Considering what you've learned in this lab activity, what structure do you think scientists use to classify fungi?

_____

10. Why is it important to classify fungi?

_____

_____

_____

11. How do the study and classification of fungi help fulfill the Creation Mandate?

_____

_____

_____

_____

_____

GOING FURTHER

Most of the imperfect fungi are plant parasites, but some are parasites of humans. Although they cause few if any fatalities, they are still very annoying. Millions of dollars are spent each year on antifungal powders, creams, and sprays.

M Two common fungal ailments are athlete's foot and ringworm. Research each of these conditions.

12. Describe athlete's foot and its common symptoms. Explain how infections spread, how they can be prevented, and how they can be cured.

_____

_____

_____

_____

_____

_____

_____

13. Describe ringworm and its common symptoms. Explain how infections spread, how they can be prevented, and how they can be cured.

_____

_____

_____

_____

_____

_____

_____

DRAWING AREA A

DRAWING AREA B

DRAWING AREA C

14A LAB

Name That Plant

Identifying Plants

Biologist Nathan Klaus was hiking near Pine Mountain, Georgia, when he came upon an amazing find. Next to the trail were six very special trees—all American chestnuts, a species on the verge of extinction. The tallest tree in this newly discovered stand is the southernmost fruit-bearing American chestnut in the world. There are only a few stands left in the entire country, though over a century ago they made up about a quarter of all the trees in the Appalachian Mountains. American chestnuts were a key species in the Appalachian food web, and Appalachian families depended on the chestnut's decay-resistant wood and abundant yields of edible nuts. But in the early 1900s a fungal disease began to wipe out the population.

How did Nathan Klaus identify these trees among all the different types in the forest? American chestnuts are identified by their flowers, fruit, and leaves. Even the shape of a tree and the color and texture of its bark can help identify it.

Now you get to try your hand at identifying some plants that may be as close as your own backyard! It's one thing to see photographs or drawings of plants but quite another to recognize actual specimens. In this lab activity you will explore plant diversity in a specific location by finding and identifying some specimens.

American chestnut

What kinds of plants live in a particular ecosystem?

Equipment

camera	hand lens
garden gloves	small notebook
garden shears	small plastic bags
field guide	permanent marker

QUESTIONS

- How can I tell what kinds of plants live in an ecosystem?

- How can I find out the names of those plants?

- Why do scientists monitor the plants in an ecosystem?

PROCEDURE

Collecting Specimens

Every scientist needs to record his observations, especially when working in the field. Observations are data, and it is important to write down information about what you observe so that you can remember it later. Sometimes the tiniest pieces of information can lead to discoveries in science. Let's set up your own personal field journal.

A Decide where you are going to observe. Obtain permission from the owner to be on the property, to take pictures, and to collect samples of the plants.

B Write down in your notebook the location you have chosen in which to observe plants. Write down the date, the time of day that you are making observations, and the people in your group. Write down your observations about the weather, wind, and soil conditions, along with the local air temperature.

1. How do weather and climate affect the plants in an ecosystem?

_____

_____

_____

_____

C Take a picture of the place where you are observing plants. Leave a spot in your journal to paste the picture. Write a brief description of the area.

Observations to note in your journal could include the color of the plant's flowers (if you don't collect any), the height of the plant, its location near a body of water, the presence of any insects on the plant, or any discoloration in the plant. Also note the geography of the area, including the presence of rocks or underbrush, or whether the ground is level or sloped.

D Make sure that you put on gloves before handling plants.

E Choose at least ten plants that you want to observe. Take a picture of the place where you collect each sample. Collect samples of each plant and place each in a plastic bag that has been labeled with a number using a permanent marker.

F Write down each specimen's number and identify it as best you can while in the field. Be sure to look at the whole plant, not just at the leaf, to identify each plant. Write the name and bag number in your field journal.

Leave lots of space for additional notes on the specimens that you identify. Use an app, a field guide, or an internet source to guide you.

A page from the field journal that William Clark kept while on the Lewis and Clark expedition

G Make notes in your journal about the specific place where you collected each of your plant samples.

2. What source(s) did you consult to identify your plants?

_____

_____

_____

These are just preliminary field identifications. Later on you'll be examining these samples again to make sure that you've identified your specimens correctly.

3. Why did you choose the plants that you did?

_____

_____

_____

_____

_____

4. What characteristics of your specimens tell you that they are all plants?

_____

_____

_____

_____

_____

_____

Verifying Specimens

Now take your specimens to a place where you can observe them more closely.

H Look at each of the specimens that you collected. Fill in Table 1 with the information given below for each specimen. You will be graded for accuracy, so be careful!

» common name

» scientific name (at least the genus)

» leaf description (needles, scales, parallel or netted venation, and simple, pinnately compound, bipinnately compound, or palmate shape)

» classification (seedless vascular, nonvascular, gymnosperm, monocot angiosperm, or dicot angiosperm)

I Confirm that your identifications are correct by comparing your specimens to the descriptions in your field guide, app, or other source. Also check a range map to make sure that the plant you've identified is found in your area. If it is not, you may want to reconsider your identification.

J Remove your gloves and wash your hands.

MONOCOT OR DICOT?

If you have trouble knowing whether your plant is a monocot or dicot angiosperm, look at the leaves. If the leaves have parallel venation, the plant is a monocot. If the leaves have netted venation, the plant is a dicot. Also, if your plant has a woody stem, it is a dicot.

dicot leaf

monocot leaf

CONCLUSION

5. After closer examination, you likely changed some of the names that you assigned to your specimens. Explain why.

_____

_____

_____

6. Which category did most of your plant specimens fall into—nonvascular, seedless vascular, gymnosperm, monocot angiosperm, or dicot angiosperm?

_____

7. Suggest a reason why there are more plants from this category.

_____

_____

_____

8. Why is it important to be able to identify and classify plants?

_____

_____

_____

_____

_____

GOING FURTHER

9. Suggest one way that scientists can monitor a population of plants like the American chestnut and help them make a comeback.

_____

_____

_____

_____

_____

An additional problem that plagues some ecosystems occurs when plants are introduced from another area and begin to take over. These invasive species crowd out the local plant species, which can make the area's biodiversity plummet.

10. Do some research to find an invasive plant species in your area. Write the common name and any recommendations that conservationists and ecologists give for controlling this plant.

_____

_____

_____

11. Write a description of how you would identify this plant if you were to see it in a local ecosystem.

_____

_____

_____

12. Why is it important to monitor plant populations, both those that are struggling and those that are thriving to such an extent that they endanger the health of an ecosystem?

_____

_____

_____

TABLE 1 Identifying Plants

Specimen Number	Common Name	Scientific Name	Leaf Description	Classification

14B LAB

A Fruitful Lab

Exploring Flowers, Fruits, and Seeds

Have you ever thought about the difference between a fruit and a vegetable? Maybe you're thinking, "Fruit tastes good, and vegetables don't!" What you might not know is that green beans and cucumbers, commonly thought of as vegetables, are actually fruits. Fruits come from the developed ovaries of flowers, while vegetables come from other parts of a plant—leaves, stems, roots, and even flowers.

Flowers are the sexually reproductive parts of angiosperms. There are many types of flowers, but they all have certain characteristics in common. When flowers are pollinated, they produce fruit, which in turn produces seeds. In this lab activity you'll explore the different types of flowers, fruits, and seeds.

What kinds of structures are usually found in different kinds of flowers, fruits, and seeds?

Equipment

scalpel
hand lens
large kitchen knife
single-edged razor blade

fresh flower specimens
fresh fruit specimens
seed specimens

QUESTIONS

- How are flowers similar and different?
- What do all fruits have in common?
- What do the seeds of monocots and dicots look like?

PROCEDURE

Observing Flowers

Not all flowers are variations of the rose, lily, or daisy. Many flowers lack showy petals, and many have very unusual structures, but most of them share the same basic floral parts.

A Using your scalpel, carefully dissect a fresh flower to see the various internal structures. The best method of dissection is to slice through the ovary area from the top down. Examine the structures with a hand lens if necessary.

B Using Drawing Area A, draw a longitudinal section of your flower. Label all the parts of a flower that you learned about in your textbook.

If a flower has sepals, petals, and at least one stamen and one carpel, it is called a *complete flower*. Flowers lacking any of these structures are called *incomplete flowers*.

1. Is your flower complete or incomplete? Explain.

_____

_____

2. Is your flower male, female, or both? Explain.

_____

_____

_____

_____

Holly trees have imperfect flowers, and each tree bears only male or female flowers. For holly trees to produce berries, there needs to be a male and female tree near one another. These kinds of plants are described as *dioecious*.

Flowers that have either male or female structures are called *imperfect flowers*. Flowers with both male and female structures are *perfect flowers*.

3. Is your flower perfect or imperfect? Explain.

_____

_____

_____

_____

Some flowers are actually several flowers held in one receptacle. These kinds of flowers are called *composite flowers*.

4. Is your flower composite or not composite? Explain.

_____

_____

_____

_____

In Chapter 14 you have learned about the differences between the leaves, stems, roots, and seeds of monocots and dicots. But how can you know whether a plant is monocot or dicot by looking at its flowers? Monocots have petals or sepals in groups of three, while dicots have petals or sepals in groups of four or five.

Trilliums are considered monocots because their sprouts have one cotyledon. You could know that trilliums are monocots by looking at their flowers—they have three petals.

5. Is your flower from a monocot or a dicot plant? Explain.

_____

_____

_____

_____

superior ovary   half-inferior ovary   inferior ovary

Now look at the ovary of your flower (if it has one). If the ovary is above the sepals, it is a *superior ovary*. If it is halfway past the base of the sepals, it is a *half-inferior ovary*. If it is below the sepals, it is an *inferior ovary*. If your flower doesn't have an ovary, skip Question 6.

6. Categorize your flower's ovary as superior, half-inferior, or inferior. Explain your classification.

_____

_____

_____

_____

Observing Fruits

A fruit is a ripened ovary. Although many fruits are like the familiar apple and orange, many are quite different. Fruits have been classified into groups on the basis of the way their structures develop.

C Using a scalpel (or large kitchen knife), dissect the fruits that your teacher gives you.

You are working with a few different kinds of fruit. These include a pome, a drupe, a true berry, a modified berry, and a legume.

D Using Drawing Area B, draw a longitudinal section of your fruit.

E Use the Key to Common Fruit Types on pages 166–67 to help identify your fruit types. Record your classifications in Table 1.

7. Describe some of the variation that you observed in your fruits.

_____

_____

_____

_____

8. Suggest reasons for the differences in fruit, especially as it relates to the differences in flowers.

_____

_____

_____

_____

Observing Seeds

The three basic parts of a seed are the plant embryo, stored food, and seed coat. The diversity of these structures found in different plants, however, is almost as wide as the diversity found in flowers and fruits.

monocot corn seed

dicot bean seed

F Using a hand lens, observe the exterior of your seeds. Draw your seeds in Drawing Area C, being careful to identify each one by its fruit.

G On your drawing, label the following parts of your seeds.

» hilum—the point where the seed is attached to the ovary

» seed coat—the hard, protective covering of the seed

H Using a scalpel (or large kitchen knife), dissect the seeds by cutting them in half. It may take some effort to open some seeds.

I Draw your dissected seeds in Drawing Area D, being careful to identify each one by its fruit.

9. Describe some of the variation that you observed in your seeds.

_____

_____

_____

_____

GOING FURTHER

10. Observe some of the other flower, fruit, and seed specimens. What did you observe in these other specimens that you didn't see in your original observations?

_____

_____

_____

11. Why is it important for people to study the different kinds of flowers, fruits, and seeds?

_____

_____

_____

_____

_____

You've learned about some of the different types of flowers, fruits, and seeds in this lab activity.

J Do some research on the different types of vegetables.

12. Name some of the different categories of vegetables and give examples of each.

_____

_____

_____

_____

_____

_____

_____

_____

_____

_____

Key to Common Fruit Types

1. (a) Single ovary that may have one or more chambers for ovules, usually without other floral parts . . . . go to 3

    (b) Collection of ovaries, usually with other floral parts . . . . go to 2

COMPOUND FRUITS

2. (a) Several separate ovaries of a single flower that ripen individually, usually on an enlarged receptacle . . . . aggregate fruit

    (b) Several ovaries from separate flowers that ripen fused together, usually on an enlarged receptacle . . . . multiple fruit

3. (a) Fruit dry at maturity . . . . go to 4

    (b) Fruit fleshy at maturity . . . . go to 10

DRY SIMPLE FRUITS

4. (a) Fruit open when ripe . . . . go to 5

    (b) Fruit closed when ripe . . . . go to 7

5. (a) Ovary wall thin, single-chambered ovary with many seeds; opens along one or two sides when ripe . . . . go to 6

    (b) Ovary wall thin, multiple-chambered ovary, each chamber with many seeds; opens when ripe . . . . capsule

6. (a) Opens along one side . . . . follicle

    (b) Opens along two sides . . . . pod

7. (a) Fruit with thin wing formed by ovary wall . . . . samara

    (b) Fruit without wing . . . . go to 8

8. (a) Thick, hard, woody ovary wall enclosing a single seed . . . . nut

    (b) Thin ovary wall . . . . go to 9

A fruit with a winged seed like a propeller blade is called a *samara*.

▼

9. (a) Ovary wall fastened to a single seed .... grain

(b) Ovary wall separated from a single seed .... achene

FLESHY SIMPLE FRUITS

10. (a) Fleshy portion develops from receptacle enlargement;
ovary forms leathery core with seeds .... pome

(b) Ovary fleshy .... go to 11

11. (a) Ovary two-layered: outer layer fleshy and inner
layer forming hard woody stone or pit, usually
enclosing one seed .... drupe

(b) Entire ovary fleshy .... go to 12

12. (a) Thin-skinned fruit with divided ovary, usually with
each section containing seeds .... true berry

(b) Thick, tough-skinned, with divided ovary, usually
with each section containing seeds .... modified berry

TABLE 1 Identifying Fruit Types

Name of Fruit	Type of Fruit

DRAWING AREA A

DRAWING AREA B

_____

DRAWING AREA C

DRAWING AREA D

15A LAB

Plant Processes

Investigating Plant Hormones and Ripening

Chances are, you've eaten a banana or two in your life. It's equally likely that you don't live anywhere near where those bananas were grown. Bananas grow in the tropics, where year-round temperatures are high and rainfall is abundant. You probably also know from experience that ripe bananas are soft and bruise easily, so shipping them when ripe is probably not a good idea. So how is it that you can find a ripe banana from Costa Rica in a supermarket in Cedar Rapids on just about any given day of the year?

To peel away the mystery of bananas, let's start by looking at how a banana changes during the ripening process.

Why can some fruits be harvested while unripe but later sold in ripened form?

Equipment

hot water bath
laboratory spatula
petri dishes (4)
disposable plastic pipette
test tubes (2)

graduated cylinder, 10 mL
unripe banana purée
iodine solution
Benedict's solution
ripe banana purée

nitrile gloves
goggles
laboratory apron

PROCEDURE

Testing an Unripe Banana

Iodine can be used to test for the presence of starches, which turn a dark blue color when exposed to iodine. Benedict's solution is used to test for the presence of simple sugars. It turns green or yellow in low sugar concentrations; orange or red results indicate higher sugar concentrations.

A Obtain a small sample of unripe banana purée from your teacher.

B Scoop a small portion of the sample into a petri dish. Use a pipette to apply a few drops of iodine solution to this portion.

1. Describe what you observe when you apply iodine to the unripe banana.

_____

2. What does this result indicate about the presence of starch in an unripe banana?

_____

C Now place 5 mL of the remaining unripe banana purée into a test tube.

D Place 5 mL of Benedict's solution into the test tube. Shake the tube well and place it in the hot water bath provided by your teacher.

> **QUESTIONS**
> - What is the difference between an unripe banana and a ripe one?
> - How do bananas ripen?
> - What role does ethylene play in the processing of bananas prior to sale?

E Allow the tube to sit in the bath for ten minutes.

3. Describe what you observe in the tube at the end of the ten minutes.

_____

4. What does this result indicate about the concentration of simple sugars in an unripe banana?

_____

Recall from Lab 7B that starch can be converted to simple sugars.

5. Using the results from testing an unripe banana for starch and simple sugars, coupled with what you learned from Lab 7B, make a prediction about the levels of starch and simple sugars that you would expect to find in a ripe banana.

_____

_____

Testing a Ripe Banana

F Repeat Steps A through E of the previous section using a sample of ripened banana purée.

6. Describe what you observe when you apply iodine to the ripened banana purée.

_____

7. What does this result indicate about the presence of starch in a ripe banana when compared to that in an unripe one?

_____

8. Describe what you observe in the tube of ripened banana purée at the conclusion of the Benedict's solution test.

_____

9. What does this result indicate about the presence of simple sugars in a ripe banana when compared to that in an unripe one?

_____

_____

10. In Lab 7B the conversion of starch to glucose was due to fermentation by yeast. Using your results from that lab activity and your personal knowledge of ripening bananas, what do you think suggests that bananas ripening is _not_ due to yeast fermentation?

_____

_____

_____

11. Summarize what you have learned so far about the difference in starch and sugar content between unripe and ripe bananas.

_____

_____

_____

_____

_____

Thinking about Ripening

Let's get back to the question of how your supermarket can offer ripe bananas for sale. Obviously, they can't be shipped while ripe since they would get damaged during the trip. You'll need to do a little sleuthing to examine this issue!

G Research ripening fruit and climacteric fruit.

12. What is a climacteric fruit?

_____

13. Are bananas climacteric or non-climacteric?

_____

14. What significance does the answer to Question 13 have for the harvest and shipment of bananas?

_____

_____

H Now do an internet search on bananas. Look for information on how bananas are harvested, shipped, and prepared for market.

15. Write a summary of what you learned.

_____

_____

_____

I Finally, do an internet search on degreening.

16. What is degreening?

_____

_____

_____

17. What do degreening and the processing of bananas have in common?

_____

_____

18. What is the major difference between degreening and the processing of bananas? (*Hint:* Recall the difference between climacteric and non-climacteric fruits.)

_____

_____

GOING FURTHER

Techniques for accelerating the ripening of fruit have been known since ancient times, but the biological mechanisms that control ripening have only recently been discovered. Ethylene's effects on plant growth were first described in 1901, and it wasn't until 1934 that ethylene was determined to be a plant hormone. The process by which plants produce ethylene wasn't completely described until the 1980s by Dr. Shang Fa Yang at the University of California at Davis. The ethylene production process in plants is called the *Yang Cycle* in his honor. Similar to cellular respiration or photosynthesis, the Yang Cycle is a complex series of chemical reactions involving intermediate products and requiring specialized enzymes. Regulator genes turn the cycle on or off, so that ethylene is produced only when the plant needs it. One popular science magazine says that how such a complex chemical pathway could have arisen by mutation and natural selection continues to puzzle evolutionary scientists.

19. What features of the Yang Cycle pose problems for the evolutionary model?

_____

_____

_____

_____

_____

20. How do those same features strengthen the position of the creation model?

_____

_____

_____

_____

_____

15B LAB

Too Salty?

Experimentation and the Flood

When irrigation water seeps through cropland, dissolved mineral salts in the water are left behind, a process known as *salinization*. It is more difficult for crops to grow when the soil has an abnormally high salt content, and that problem can have a significant impact on a culture. For example, as salt built up over centuries in the soil of ancient Mesopotamia, entire civilizations collapsed.

Receding floodwaters from Noah's Flood likely left salt water on the land. After drying, the land would have had a higher-than-normal salt content. How did plants grow and reestablish ecosystems after the Flood?

?

If seawater from Noah's Flood was salty, how did plants survive the Flood?

Equipment

containers to soak seeds (2)

containers to hold seeds during germination (3)

spray bottle

paper sack

plant seeds

paper towels

seawater

distilled water

PROCEDURE

Background

This question was partially addressed by the paper "Seed Germination, Sea Water, and Plant Survival in the Great Flood," by George F. Howe in *Creation Research Society Quarterly*, Vol. 5, No. 3 (December 1968), pp. 105–12.

Howe's paper went through a *peer review* process before it was published, a standard practice in scientific research. When an article is peer reviewed, it is examined by other scientists working in the same field before it is published. They read the article and make suggestions for improvement. The peer review process raises the quality of the article.

QUESTIONS

- How does salt water affect seed germination?
- How can experimentation aid our thinking about biblical issues like the Flood?
- What is peer review?

The paper by Howe follows a common style of scientific writing. Notice the following sections of the paper.

TITLE

The title conveys the content of the paper in a concise way.

ABSTRACT

A well-written abstract allows a scientist to read the abstract and decide whether to read the rest of the paper. It also allows the scientist to come away with the major details of the paper and not have to read the paper in depth. An abstract summarizes the entire paper in one or two paragraphs and is often limited to 200–300 words. It usually is the last part of the paper to be written, even though it is found next to the title. An abstract will typically answer the following questions.

- » What was the purpose of the study?
- » What was the experimental design of the study?
- » What were the major findings of the study?
- » What were the conclusions of the study?

INTRODUCTION

The introduction starts broad and then focuses the reader on the work reported in the paper. A good introduction will answer the following questions.

- » What was studied?
- » Why was this topic important?
- » What did we know about this topic before this study?
- » What new things will we learn because of this study?

MATERIALS AND METHODS

The materials and methods section will sometimes be referred to as simply "methods." The purpose of this section is to explain how the study was done. It will contain details about control groups, experimental groups, experimental techniques, procedures, and data analysis methods.

RESULTS

The results section reports data collected. This section will often have data tables and graphs.

DISCUSSION

The discussion section involves the interpretation of the results. The original research question or hypothesis is compared to the results, which are then compared to previous research. A few sentences are usually dedicated to suggestions for extending the research.

REFERENCES

The references section lists the papers and books referred to by the article.

The format for scientific writing is very efficient at presenting information. Because this format is fairly standard across all scientific disciplines and because scientists are accustomed to using and reading papers written in that style, it is relatively easy for the reader to move around in the paper depending on his level of interest.

Howe's Research

A Read the above-referenced paper by Howe. Your teacher will give you directions about how to access it.

1. What scientific principle explains why we expect that salt water will affect seed germination? Explain.

   _____

   _____

   _____

   _____

2. Why did Howe thoroughly rinse the soaked seeds before planting them?

   _____

   _____

   _____

   _____

   _____

3. This exercise is a model for the Flood and post-Flood environments. What is the counterpart in the Flood narrative for the rinsing of seeds before planting?

   _____

   _____

   _____

   _____

4. Why did Howe collect fruits with seeds and soak them in the different water mixtures rather than just soaking the seeds alone in the different water mixtures?

   _____

   _____

   _____

   _____

   _____

Howe lists four groups of seeds that he experimented on:

» fruits stored in paper sacks

» fruits soaked in seawater

» fruits soaked in seawater mixed with tap water

» fruits soaked in tap water

5. Explain why the fruits stored in paper sacks were the control groups.

_____

_____

6. Explain why fruits soaked in tap water was a necessary group for the experiment.

_____

_____

_____

_____

_____

_____

7. Why did Howe's study include a group of fruits soaked in seawater mixed with tap water?

_____

_____

_____

_____

_____

8. What conclusion did Howe draw from his experiments?

_____

_____

9. List two suggestions for future work that Howe suggested on the basis of his experiment.

_____

_____

_____

_____

_____

Seed Germination Experiment

B Set up two containers and fill one with fresh water and one with seawater.

C Place 20–25 seeds in each container and in a paper sack. Allow the seeds in the containers to soak for the time prescribed by your teacher. You will also be instructed regarding any needed maintenance of the seeds while they are soaking.

D Research the effects of salt on plant growth for background material to use in the introduction of your lab report.

E After the seeds have soaked for the specified time, remove them from the water and wash them thoroughly with distilled water.

F Place paper towels two layers thick in a container. Place the seeds from the salt water on top of the paper towels and place another two layers of paper towels over the seeds. Spray the paper towels with distilled water until they are soaked.

G Repeat Step F for the seeds from the fresh water and from the paper bag.

H Examine the seeds regularly over a couple of weeks.

I Write a lab report of your findings using the style of a scientific paper. Use Howe's paper as an example to follow. The lab report should contain the sections outlined on page 176.

10. Write a paragraph about how this study demonstrates ways in which a biblical worldview affects science.

_____

_____

_____

_____

_____

16A LAB

The Immortals Next Door

Investigating Hydras

What if there were a way to stay ever young, fit, and healthy—a real-life Fountain of Youth? Some scientists think that a certain animal, one that might not seem an obvious candidate, may hold the key to unlocking the secret of eternal youth: the lowly hydra. Hydras are common in freshwater lakes and streams in temperate zones around the world, but you may have never noticed them because of their very small size. They can sting like jellyfish, but their venom is too weak to be a danger to humans. Let's take a look at this tiny representative of the phylum Cnidaria.

How do hydras move, feed, and react to their environment?

Equipment

microscope
hand lens
petri dish
disposable plastic pipettes (3)
probe

concavity slide
cover slip
preserved slides of a budding hydra and hydra cross section
spring water

live cultures of hydras and brine shrimp or *Daphnia*
dilute acetic acid

PROCEDURE

Observing Preserved Hydras

A Observe a preserved budding hydra on low power.

B Make an outline drawing in Drawing Area A. The entire hydra will not fit into the microscope field, so you will need to move the slide several times as you draw.

C Label the tentacles, mouth, basal disk, and buds.

D Using high power, draw a cross-section view of a portion of the hydra body wall in Drawing Area B. This drawing should include internal structures and not just be an outline.

E Label as many of the cellular structures of the hydra as you can (see page 358 in your textbook).

QUESTIONS

- How do hydras move and feed?

- Do hydras respond to changes in their environment?

- Why are hydras the focus of much current research?

F Fill a clean petri dish with spring water.

G Using a pipette, draw in some water from the hydra culture provided by your teacher.

H Select a hydra and use the pipette to flush it, dislodging the specimen.

I Draw the dislodged hydra into your pipette and then flush it into your petri dish.

J Use a hand lens to observe your living hydra.

1. Describe any forms of locomotion or movement that you observe.

_____

_____

2. On the basis of what you have read about the cnidarian nervous system in your textbook, make a prediction about what kinds of responses you will be able to observe in your hydra.

_____

_____

K Very gently swirl the water around the hydra.

3. What is the hydra's reaction to the movement of water?

_____

_____

L After the hydra has recovered from the above experiment (which may take several minutes), touch your probe as gently as possible to its base.

4. What is the hydra's reaction to the probe at its base?

_____

_____

M After the hydra has recovered from the probe in Step L, touch your probe as gently as possible to one of its tentacles.

5. What is its reaction to the probe at its tentacles?

_____

_____

N After the hydra has recovered from the probe in Step M, arrange your probe so that the hydra may touch the probe of its own power.

6. What is the hydra's reaction to the probe when it touches it of its own power?

_____

_____

Observing a Feeding Hydra

O Using a pipette, place a few living *Daphnia* or brine shrimp in your petri dish near the hydra. Be careful not to add the food so rapidly that you disturb the hydra. Do not try to force-feed the hydra but keep the food within tentacle reach.

P Using a hand lens, watch the actions of your hydra carefully. Especially note any activity in the mouth region.

7. Describe the feeding process that you observe.

_____

_____

8. In what structure of the hydra does extracellular digestion take place?

_____

9. What happens to any substances that cannot be digested?

_____

Observing a Hydra's Response to Acid

Q Carefully remove your hydra and set it in a large drop of spring water placed on a concavity slide.

R Place a cover slip over the slide and observe the hydra (or sections of it) with a microscope.

S After the hydra has recovered from the transfer and you have been able to focus on the cells of its tentacles, place a small drop of the dilute acetic acid on the edge of the cover slip. Watch the hydra carefully. For the best results, one partner should put the acetic acid on the slide while the other partner observes the reaction through the microscope.

10. What did you observe as the acid reached the hydra?

_____

_____

CONCLUSION

11. How specialized to different stimuli are the responses of the hydra? In other words, is the hydra able to respond differently to various kinds of stimuli? Explain.

_____

_____

_____

_____

12. Why are the reactions of the hydra limited?

_____

_____

_____

_____

To remove your hydra, flush the slide with spring water while holding it over a separate culture container for "used" hydras. Do not return the hydra to its original culture.

GOING FURTHER

In a 1998 article in the journal *Experimental Gerontology* researcher Daniel Martinez claimed that hydras are biologically immortal. Subsequent research seems to back that claim. This phenomenon is known as *non-senescence*, meaning that the mortality rate, or rate at which individuals within a population die, does not increase with age as is typical for most other organisms. Scientists are still investigating what makes this possible.

Researching the hydra is just one of the many efforts currently underway to try to radically lengthen the human life span. There's even a name for this movement: *immortalism*.

13. According to a biblical worldview, why does physical death exist within God's creation?

_____

_____

14. John 3:16 clearly states that eternal life is possible. According to this verse, how is eternal life obtained?

_____

_____

15. In view of the truth of John 3:16, do you think that science can ever solve the problem of physical death? Explain.

_____

_____

_____

_____

_____

_____

16. Can research aimed at extending human life expectancy be compatible with a biblical worldview? Explain.

_____

_____

_____

_____

_____

_____

_____

17. Why should a Christian not be afraid of death? (See 1 Corinthians 15:55.)

_____

_____

_____

_____

_____

_____

_____

DRAWING AREA A

DRAWING AREA B

16B LAB

Fish Tank Fiend!

Observing Planarians

Al the Amateur Aquarist came home after a hard day at work, grabbed a cold soda from the refrigerator, headed for the living room, and plopped into his comfortable recliner. From there he cast his gaze onto his most prized aquarium—a 300 gallon hi-tech, heavily aquascaped beauty housing exotic plants and fish from the Amazon basin. Al had spent many months and a lot of money getting this tank just right. For Al there was no better way to unwind than to watch the dazzling shoal of cardinal tetras, the stately promenade of angelfish, the gentle swaying of the plants in the outflow of the filter, and—wait a minute. What were those little brown things slithering along the inside of the front glass panel? Al got up to get a closer look. Could they be? Oh no, they were—planarians! Argh!!

How do planarians interact with their environment?

For Al and other aquarium keepers, planarians are unsightly pests that probably hitched a ride into their tanks on bundles of plants from the local tropical fish store. But for certain scientists, planarians are a subject of intense research. Not only are planarians an important part of aquatic ecosystems, scavenging the bottom for detritus, but planarians may hold the key to a much sought-after advance in medical treatment. Let's begin by observing a live planarian.

Equipment

petri dish

disposable plastic pipettes (several)

hand lens

forceps

probe

spring water

culture of living planarians

Epsom salt

index card or small piece of poster board

raw beef liver, boiled egg yolk, or flaked or pelleted fish food

QUESTIONS

- How do planarians behave?

- Why do people study planarians?

PROCEDURE

Observing Planarian Movement

First let's look at how planarians move.

A Fill a petri dish with spring water until the bottom is just covered.

B Obtain a planarian from the culture provided by your teacher.

C Agitate the water around the planarian by making small currents with a pipette until the planarian floats around in the water, and then quickly draw it into the pipette.

D Expel the planarian into the petri dish that you have prepared before it has a chance to attach itself to the inside of the pipette.

E Give the planarian five minutes to acclimate to its dish before continuing.

F Using a hand lens, study the two ways planarians move: ciliary movement and muscular movement.

1. By observing the movement of your planarian within its dish, determine where you think its cilia are located.

_____

2. After observing the planarian moving by using its cilia, describe this type of movement.

_____

G To see muscular movement, place a few grains of Epsom salt into the water right next to the planarian. Watch carefully as the Epsom salt begins to dissolve near it.

H After you have observed the movement that results, use forceps to remove and discard the grains of salt, and then add some fresh spring water to dilute the salt.

3. Describe muscular movement in comparison to the ciliary movement that you described above.

_____

_____

_____

_____

_____

Observing Planarian Responses

Now that you've observed how planarians move, let's look at how they respond to stimuli in their environment. You should conduct each of the following tests several times to be sure that your planarian is exhibiting an actual response instead of simply making a random movement.

I With a clean probe, very lightly touch one side of the planarian.

4. Describe the planarian's response.

_____

_____

J Lightly touch one of the planarian's auricles (the side points of its head).

5. Describe the planarian's response.

_____

_____

K Lightly touch the planarian's posterior end.

6. Describe the planarian's response.

_____

_____

7. What do the planarian's responses all have in common?

_____

_____

Now observe your planarian's response to current.

L Fill a clean pipette with water and slowly force the water out to produce a current. Direct the current to one side of the planarian in the petri dish. (Have ready several pipettes full of water.)

8. Describe the planarian's response.

_____

_____

M Direct the current toward the side of the planarian's head.

9. Describe the planarian's response.

_____

_____

N Direct the current toward the posterior end of the planarian from directly behind it.

10. Describe the planarian's response.

_____

_____

O Next, observe your planarian's response to light. Cover half of the petri dish with an index card, small piece of poster board, or similar, so that half the dish is shaded. Make sure that your planarian is in the lighted half of the dish.

11. Describe the planarian's response to light.

_____

_____

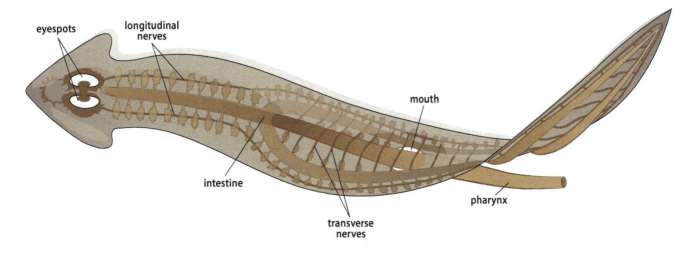

eyespots

longitudinal nerves

mouth

intestine

pharynx

transverse nerves

Observing Planarian Feeding

Refer to the sketch above to answer Questions 12–15.

12. Describe the mouth's location on the body.

_____

_____

13. Describe the appearance of the pharynx.

_____

_____

14. Describe the area within the planarian where digestion begins. Be sure to use the proper name for this structure.

_____

_____

_____

_____

15. After the structure described above completes preliminary digestion, what must happen to any matter that cannot be digested? Explain.

_____

_____

_____

_____

P Place a small piece of food near your planarian and observe its behavior.

16. Describe the planarian's response to food.

_____

_____

_____

_____

17. How does the planarian's range of responses (how specific its responses are) compare to that of the hydra?

_____

_____

_____

_____

GOING FURTHER

There are a variety of ways that our flustered aquarist friend Al could get rid of his planarian pests, but one method he should probably *not* consider is catching the worms, chopping them up, and feeding them back to his fish. They might enjoy the treat, but the problem with this method is that every piece of uneaten worm has the potential to grow into an entire new worm. In fact, in one study planarians grew back from pieces as small as *1/279th* of the original worm! This amazing ability to regrow from pieces is called *regeneration*, and although other animals can regenerate to one degree or another, planarians are particularly adept at it.

Q Do an internet search for planarian regeneration. After looking at a few sources, answer the following questions.

18. What specific aspects of regeneration in planarians do scientists find particularly intriguing?

_____

_____

_____

_____

_____

19. What is the current focus of research on planarian regeneration?

_____

_____

_____

_____

20. Why is an understanding of the mechanism of planarian regeneration of interest to human medical research?

_____

_____

_____

_____

_____

21. Why does the subject of regeneration of human tissue pose ethical dilemmas?

_____

_____

_____

_____

_____

22. Is stem cell research compatible with a biblical worldview?

_____

_____

_____

_____

_____

_____

17A LAB

Take a Crack at Crayfish

Dissecting a Crayfish

Crustaceans are a diverse group of animals. Crayfish are typical crustaceans that look like miniature lobsters. As you dissect one today, keep in mind that the structures you observe can be found in many other crustaceans.

What are the structures in a typical crustacean?

Equipment

dissection pan

dissection kit

camera

culture dish

flower head pins (8)

plastic bag

fine-tip permanent marker

preserved crayfish

glue, rubber cement, or tape

laboratory apron

nitrile gloves

goggles

PROCEDURE

The External Anatomy of the Crayfish

A Place your crayfish in a dissection pan and carefully observe it. You will notice that its body is made of multiple segments, each bearing a pair of appendages. Count the pairs of appendages on your crayfish. As you read the names and descriptions of the body structures in the key on page 193, locate them on your specimen.

B Write the number associated with each structure shown on page 192 with the structure's name on the key.

QUESTIONS

- What arthropod characteristics can be seen in a crayfish?

- What appendages does a crayfish have?

- How are a crayfish's internal organs specialized for its lifestyle?

EXTERNAL ANATOMY of a CRAYFISH

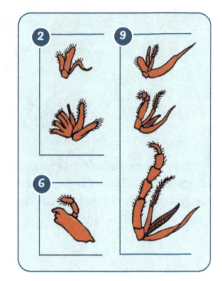

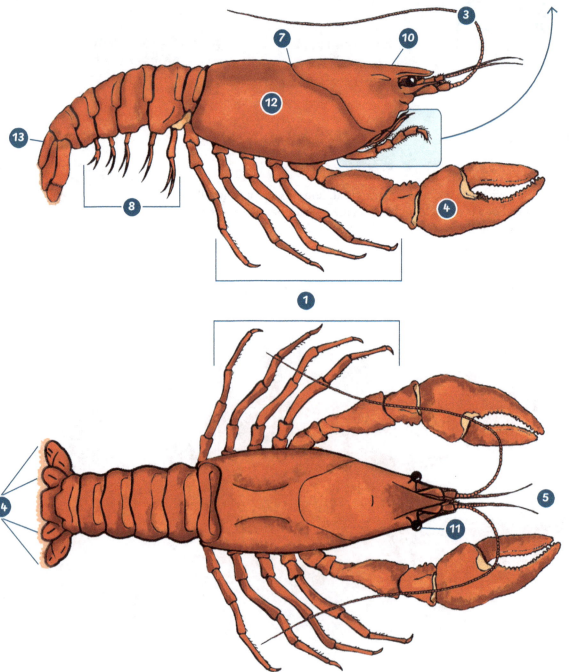

Key to the External Anatomy of a Crayfish

_____ **1.** antennules—These most anterior appendages are used in the senses of balance, taste, and touch.

_____ **2.** antennae—Longer than the nearby antennules, these are used in the senses of taste and touch.

_____ **3.** mandible—Also called true jaws and located just posterior to the antennae, these small, hard coverings of the mouth pulverize food.

_____ **4.** maxillae (two pairs)—Just posterior to the mandibles, the maxillae assist in chewing.

_____ **5.** maxillipeds (three pairs)—Just posterior to the maxillae, the highly branched maxillipeds are used to hold food in place.

_____ **6.** compound eyes—These are technically not appendages.

_____ **7.** carapace—This single section of exoskeleton covers the cephalothorax.

_____ **8.** rostrum—This extension of the carapace forms a horny beak between the eyes.

_____ **9.** cervical groove—This depression marks the division between the head and the thorax.

_____ **10.** chelipeds—These large claws are used for protection and capturing food.

_____ **11.** walking legs (four pairs)—Notice that the legs are different from one another.

_____ **12.** swimmerets (five pairs)—These abdominal appendages move water over the crayfish's gills and function in reproduction.

_____ **13.** uropods—These paired appendages grow from the posterior segment of the abdomen.

_____ **14.** telson—This structure also grows from the posterior segment.

C Examine the cephalothorax of your preserved crayfish.

15. How many segments does a crayfish's cephalothorax have?

_____

16. Locate and examine the mouth parts. Human mouthparts move vertically (up and down). In what direction do the mouthparts of a crayfish move?

_____

17. How do the walking legs differ from one another?

_____

18. Which of the walking legs, if any, have pincers?

_____

D Examine the abdomen of the crayfish.

19. How many segments does the abdomen have?

_____

20. Locate the anal opening on the ventral side of the crayfish. On which segment is it located?

_____

In males, the most anterior swimmerets are enlarged and point anteriorly. In the female, the anterior swimmerets are greatly reduced in size.

21. What is the sex of your crayfish?

_____

22. The telson and the uropods form a powerful tail fan. How would the crayfish use these?

_____

_____

The Internal Anatomy of the Crayfish

Carefully read the following directions for each body part before starting your dissection. The following procedures must be done in order. You will be expected to know the functions and locations of the organs that have been italicized.

THE BODY CAVITY

Place your animal in the dissection pan with its dorsal side up.

E Carefully insert the point of the scissors under the dorsal surface of the carapace at the posterior edge of the cephalothorax. Cut anteriorly along the midline of the body to the rostrum.

F Reposition the scissors just behind the eyes and make a transverse cut.

G Carefully remove the two pieces of the carapace without disturbing the structures underneath.

THE RESPIRATORY STRUCTURES

H Remove a few *gills* and place them in a culture dish of water.

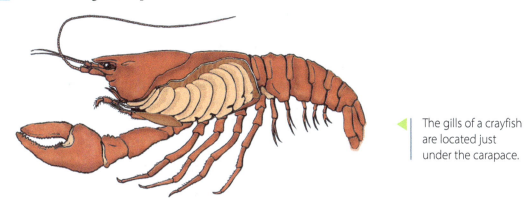

◀ The gills of a crayfish are located just under the carapace.

23. Describe their function.

_____

_____

_____

I Carefully remove the rest of the gills.

24. When a crayfish spends time on land, it holds water inside its carapace. Why would it do this?

_____

_____

_____

THE CIRCULATORY STRUCTURES

J For easier handling of the animal, remove the legs attached to the thorax.

K Carefully separate the dorsal tissues in the thorax and locate the *mid-dorsal heart*.

L Locate the main blood vessels attached to the heart.

The crayfish, like most arthropods, has an open circulatory system.

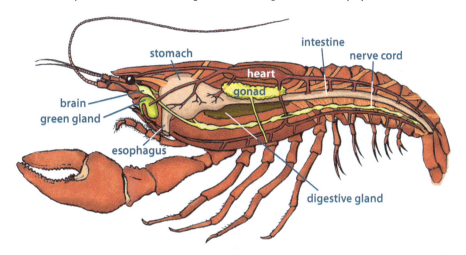

25. How does an open circulatory system work?

_____

_____

26. What force does an open circulatory system rely on that a closed circulatory system does not?

_____

27. Why would being immobilized on its back be lethal to a crayfish?

_____

_____

THE REPRODUCTIVE AND DIGESTIVE STRUCTURES

M The two light-colored masses extending along each side of the body cavity beyond the cervical groove are the _digestive glands_. Look between the digestive glands to find the reproductive structures.

» If your crayfish is a male, look for a small pair of white _testes_ and _coiled ducts_.

» If your crayfish is a female, look for a large mass of dark-colored eggs inside the _ovaries_.

28. What is the function of the digestive glands?

_____

N To expose the _intestine_, insert the point of the scissors underneath the dorsal side of the exoskeleton covering the abdomen. Cut posteriorly to the final segment.

» Open the abdominal exoskeleton along the cut. The intestine appears as a tube on the dorsal side of the abdominal muscles.

» Do not confuse the intestine with the dark-colored dorsal blood vessel.

O Trace the intestine anteriorly to the portion of the cephalothorax where the intestine joins the large, thin-walled _stomach_.

P Number your pins one to eight using a permanent marker. Insert the following pins in these organs:

» Pin 1—heart

» Pin 2—primary reproductive structure

» Pin 3—either of the digestive glands

» Pin 4—stomach

» Pin 5—intestine

Q Take a picture of your crayfish with the pins and paste it in Photo Area A.

R Remove most of the crayfish's internal organs.

» Just behind the eyes, cut the bands of muscles leading to the stomach.

» Pull the stomach posterior and cut the short esophagus located just below the stomach.

S Carefully lift out the organs all in one piece. Notice the mesenteries (connective tissue), which keep the organs all together.

THE EXCRETORY STRUCTURES

T Clean out the remaining tissue in the head to expose the *green glands* just posterior to and below the antennules. These soft, small, and only slightly green organs filter nitrogenous wastes from the crayfish's body, similar to how kidneys function in a vertebrate.

U Being careful, look for the small, sac-like bladder, which is connected to the green glands.

THE NERVOUS SYSTEM

V At the front of the head cavity, between the eyes, note the *brain*, a tiny mass of white tissue.

W Trace the nerves that go from the brain to the antennae and the eyes.

X Trace the *nerve cord* from the brain to the abdomen by cutting the hard tissue on the floor of the thorax with a scalpel.

» Spread the abdomen apart and pull out the large muscles.

» The nerve cord should now be exposed on the ventral side of the abdomen.

Y Insert the following pins in these organs:

» Pin 6—either of the green glands

» Pin 7—brain

» Pin 8—nerve cord

Z Take a picture of your crayfish with the pins and paste it in Photo Area B.

29. What are the swollen portions of the nerve cord called?

_____

30. Why is it an advantage for the crayfish to have its nerve cord on the ventral side rather than on the dorsal side, as it is in humans?

_____

_____

PHOTO AREA A

PHOTO AREA B

17B LAB

Cricket Caper

Inquiring into House Crickets

The humble house cricket has had a checkered relationship with humans. Some people have found them inspiring, keeping them as pets or elevating them to respectable status in children's stories such as George Selden's *The Cricket in Times Square*. Other people know them as household pests, worthy only of a scornful swipe of a broom. Somewhere along the way someone discovered that crickets were an easily cultured form of live food for more desirable pets such as birds and reptiles. Even people themselves eat crickets in certain parts of the world—in Thailand they are deep-fried and considered a tasty snack. Don't scoff—they're actually a good source of protein!

But some scientists, like Dr. Rajat Mittal at Johns Hopkins University, are interested in crickets for reasons other than their literary merit or culinary qualities. Mittal and his students are practitioners of "bioinspired engineering." They use high-speed photography to study the jumping motion of spider crickets in an attempt to understand how the crickets' spindly legs power their amazing leaps—the equivalent of a human jumping the length of a football field. They hope to apply what they learn to the manufacture of tiny jumping robots that can navigate uneven ground. It is your task in this lab activity to think of an aspect of cricket behavior that you can investigate.

What can I learn from studying cricket behavior?

Equipment
crickets

QUESTIONS

- What are some cricket behaviors that can be investigated?

- What kinds of experiments can be designed to test those behaviors?

PROCEDURE

Plan and Write Scientific Questions

First, you'll need a hypothesis to test. There are many aspects of cricket behavior that you could investigate and different methods that you could use to test those behaviors. To help get you started, think about three broad categories of behaviors.

» How do crickets respond to their environment?

» How do crickets interact with other crickets?

» How do crickets interact with other organisms?

Brainstorm with your group to come up with a list of possible questions to investigate. Consider which ones interest you the most. You'll need to take into account whether you have the necessary equipment on hand (or can easily obtain it) and enough time to adequately test your possible hypotheses. Narrow down your list of questions to the one you will investigate.

1. What question will you investigate regarding cricket behavior?

Scientific inquiry requires testable statements related to the question being investigated. These statements are called *hypotheses*. Think about a testable hypothesis that you can use to help you answer your question about cricket behavior. Have your teacher approve your hypothesis.

2. What is the hypothesis that you will test?

_____

_____

Designing and Conducting Scientific Investigations

Next, you will need to design an experiment to test your hypothesis. Get approval for your experimental design from your teacher before beginning your experiment.

3. Briefly describe your experimental setup.

_____

_____

_____

4. What will be your control group?

_____

_____

5. What will be your experimental group?

_____

_____

6. What data will you be collecting?

_____

_____

7. How will you analyze your data?

_____

_____

Set up and run your experiment. Make sure to run several trials to solidify support for your conclusions.

Developing Models

Once you have collected your data, you will need to analyze it by using mathematical tools, such as finding and comparing averages or creating graphs.

Scientific Argumentation

In Lab 15B you learned about scientific writing. Recall that a scientific paper has distinct sections, including an abstract, an introduction, a discussion of materials and methods, the results, a discussion of the meaning of the results (including suggestions for further research), and a list of references. Your teacher will instruct you on whether to report your findings in a formal lab report (a type of scientific paper), an oral report, or a slide presentation. As part of your report or presentation, describe how the behavior you observed gives evidence of God's wisdom in providing for the cricket's needs.

18A LAB

Something Fishy Going On

Observing Bony Fish

In the sandy depths of Lake Malawi in East Africa there lies a fish—literally. The locals call it *kalingono*—the sleeper. Only the fish isn't sleeping. It's hunting. As it lies on its side on the sand, the splotchy brown and white sleeper looks remarkably like a rotting corpse. When a smaller fish comes near to investigate—wham! The sleeper lunges and swallows the unsuspecting prey whole.

With over 30,000 described species so far, as many or more than all other vertebrates combined, you'd expect the many varieties of fishes in the world to exhibit a wide range of behaviors. Some, like the sleeper's, are truly bizarre. In this lab activity you'll take a closer look at the external form and behavior of a fish of your choice.

Nimbochromis livingstonii, **the sleeper**

How do fish behave?

?

Equipment

live fish to observe
notepad and pencil

QUESTIONS

• What are the key external features of a fish?

• How does a fish move and function in its environment?

• How does a fish respond to its environment?

PROCEDURE

First, you'll need a live fish to observe. It can be your own fish, a neighbor's fish, a fish at a pet store, or a fish in a public aquarium. Whatever fish you choose, it is important that the fish is in a setting that makes it feel comfortable so that it will behave as naturally as possible. A fish in a harshly lit, overcrowded aquarium with no hiding places will be stressed and not look or act its best.

Observing any kind of wildlife takes time. That means that you, the observer, need to give some thought to your own comfort as well. You'll need to observe your fish for an extended period of time without a lot of motion. Once you have chosen a fish to observe and have gotten yourself comfortably positioned, give the fish about five minutes to become acclimated to your presence. Some fish held in captivity quickly learn to associate humans with food; they need time to figure out that you're not there to feed them and go back about their business. Other fish are shy and retiring and need a few minutes to be sure that you're not a danger to them.

When your fish has become acclimated to you and you are ready to start recording observations, try to be slow and deliberate with your motions—don't make sudden movements. If the fish is startled and makes a dash for cover, give it a few minutes to relax again before continuing.

Form and Features

1. What is the common name for your fish?

_____

2. What is the scientific name of your fish?

_____

A Observe your fish from several angles with an eye for its overall shape and form.

3. Describe the body shape as viewed from the side.

_____

_____

4. Describe the body shape as viewed from the front.

_____

_____

5. In what ways is the fish's body form ideally suited for its aquatic environment?

_____

_____

_____

_____

B Study the fish's fins closely and indicate their number, location, and shape in the appropriate columns of Table 1 (see page 401 in your textbook for help). In the *Rays* column, indicate whether each fin has soft rays, spiny rays, both, or none. In the *Function* column, include whether each fin is used primarily for steering, stability, or propulsion.

The highly modified fins of a lionfish

6. Is the mouth of your fish superior (upturned), terminal (pointing directly ahead), or inferior (downturned)?

_____

7. What does the position of your fish's mouth suggest about its feeding strategy?

_____

_____

_____

_____

_____

C Notice the size and position of your fish's eyes.

8. What does the size of your fish's eyes suggest to you regarding the importance of vision for fish?

_____

_____

_____

9. What does the position of your fish's eyes tell you about its field of vision?

_____

_____

_____

_____

D Try to locate your fish's lateral line (see page 402 in your textbook).

10. What is this line made of, and what is its function?

_____

_____

_____

11. How does the arrangement of the lateral line enhance its role as a sensory organ?

_____

_____

_____

_____

E Watch your fish's breathing carefully.

12. What are the two primary external structures involved in fish breathing?

_____

13. Describe how these two structures work together as the fish breathes.

_____

_____

_____

_____

14. Describe any other interesting external features your fish might have.

_____

_____

_____

_____

15. Describe your fish's coloration.

_____

_____

_____

_____

F In Drawing Area A, make a sketch of your fish. Label the fins and any structures that you were able to identify in Questions 6–14.

Behavior

G Now take some time to carefully observe how your fish interacts with other fish and with its surroundings.

16. Where in the water column does your fish spend most of its time (upper, mid, lower)?

_____

17. Is your fish active, swimming unceasingly about its enclosure? Or does it limit its movement to one particular area? Describe your fish's general activity.

_____

_____

18. Does your fish prefer open water, or does it prefer a sheltered area? How does your fish utilize the space available to it?

_____

_____

_____

19. How are your answers to Questions 17 and 18 related?

_____

_____

_____

20. How does your fish propel itself forward?

_____

_____

_____

21. Can your fish move backward? If so, how?

_____

_____

22. Compare your fish with other fish in the tank. Does there appear to be a relationship between the activity of your fish described in Questions 17 and 18 and the location of its pectoral and pelvic fins? Explain.

_____

_____

_____

_____

_____

_____

23. Does your fish spend its time in the company of other fish of the same species (schooling), or does it keep to itself?

_____

Head down, fins flared. This tiny fish is telling its tankmates, "Back off!"

24. Does your fish exhibit any aggression toward others of its kind? If so, describe its actions.

_____

_____

_____

_____

25. Does your fish show signs of aggression to other species of fish in the tank? If so, describe its actions.

_____

_____

_____

_____

26. Does your fish show any signs of territoriality (having a territory)? If so, describe its actions.

_____

_____

_____

_____

27. Describe any other behaviors that you observe during your viewing session.

_____

_____

_____

_____

GOING FURTHER

Fishkeeping is big business. Some sources have estimated that roughly 10% of all American households have an aquarium, and the hobby is popular in Europe and Asia as well. Several billion dollars are spent each year on aquariums, fish, and supplies. Large quantities of aquarium fish are bred on farms, but many are also harvested from the wild. This is particularly true of saltwater reef fishes. Do an internet search for "ethics and tropical fish." After reading some articles, answer the following questions.

28. What are some of the ethical issues pertaining to the harvesting and keeping of tropical fish?

_____

_____

_____

_____

_____

_____

29. Can the hobby of keeping tropical fish be reconciled with a biblical worldview? Explain.

_____

_____

_____

_____

_____

_____

Tropical fish farms provide an alternative to wild-caught fish.

TABLE 1 Fish Fins

NAME	NUMBER	LOCATION	SHAPE	RAYS	FUNCTION
Pectoral Fin					
Pelvic Fin					
Dorsal Fin					
Anal Fin					
Adipose Fin					
Caudal Fin					

DRAWING AREA A

18B LAB

Reptile Repasts

Inquiring into Reptile Methods of Locating Prey

Lizards eat a variety of foods, and their methods of finding food depend on the species. In this lab activity you will design an experiment to test a particular species of lizard's method of hunting.

Equipment
terrarium
lizard

PROCEDURE

Plan and Write Scientific Questions

A Think about how your lizard might use its senses to find food. Brainstorm with your group to come up with a list of possible questions to investigate. Consider which areas of investigation interest you the most. You'll need to consider whether you have the necessary equipment on hand (or can easily obtain it) and enough time to adequately test your possible hypotheses. Narrow down your list of questions to the one that you will investigate.

1. Which of your lizard's senses will you be testing?

_____

2. Write a hypothesis about how your lizard uses the sense you've selected to find its prey.

_____

_____

Designing and Conducting Scientific Investigations

You now need to design an experiment to test your hypothesis. Get approval for your experimental design from your teacher before beginning your experiment.

3. Briefly describe your experimental setup.

_____

_____

_____

How do lizards detect their prey?

QUESTIONS
- How do lizards find food?
- Do lizards rely on some senses more than others when hunting?

4. What will be your control group?

_____

5. What will be your experimental group?

_____

6. What data will you be collecting?

_____

_____

_____

_____

7. How will you analyze your data?

_____

_____

_____

_____

Set up and run your experiment. Make sure to run several trials to solidify support for your conclusions.

Developing Models

Once you have collected your data, you will need to analyze it by using mathematical tools, such as finding and comparing averages or creating graphs.

Scientific Argumentation

In Lab 15B you learned about scientific writing. Recall that a scientific paper has distinct sections, including an abstract, an introduction, a discussion of materials and methods, the results, a discussion of the meaning of the results (including suggestions for further research), and a list of references. Your teacher will instruct you on whether to report your findings in a formal lab report (a type of scientific paper), an oral report, or a slide presentation. As part of your report or presentation, describe how the behavior you observed gives evidence of God's wisdom in providing for the lizard's needs.

19A LAB

Our Fine, Feathered Friends

Creating a Birding Log

Many migratory birds travel thousands of kilometers each year on their annual migrations. That's a lot of frequent flier miles! It would be impossible for any ornithologist—a scientist who studies birds—to keep a constant watch on such well-traveled birds. But happily, ornithologists have a large, dedicated, and energetic group of unpaid assistants who are only too eager to provide them with valuable data on bird numbers and movements. They're *birders*.

Amateur birdwatching, or birding, is a popular pastime for many people and an invaluable tool for scientists as well. Many birders report their bird sightings to birding organizations. Scientists use these reports to estimate the number of different bird species and their population sizes. They also provide valuable data on the ranges and migration paths of those species.

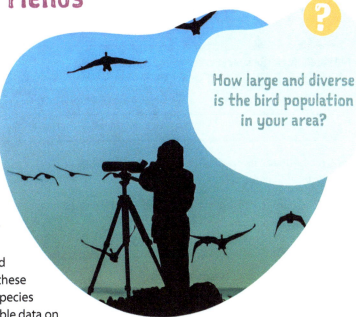

How large and diverse is the bird population in your area?

Equipment
birding app or field guide
camera
notebook

QUESTIONS
- How many birds live in your area?
- How many species of birds live in your area?
- How many individuals of each species of bird live in your area?

PROCEDURE

Your goal is to identify ten or more species of birds that live in your area.

A During the time period indicated by your teacher, look for birds in the wild. This can be in a city park, in a rural field, or anywhere else you see birds.

B When you find a bird, use a birding app or field guide to identify the bird. If you have a camera, take a picture of the bird. Record its common and scientific names, its habitat (e.g., field, stream bank, urban neighborhood), and its location in Table 1. (If available, include the street address or location name. At the very least, record the name of the county and state. If you have access to a GPS device, record the GPS coordinates of the location.)

C For each species of bird that you identify, record the number of individuals of that species that you see during your project in the *Number* column of Table 1.

D If you took pictures of your identified species, attach them to a separate sheet of paper. Add a label identifying each one by its common and scientific names.

1. How many species of birds did you identify?

_____

2. What was your most numerous species?

_____

3. What was your least numerous species?

_____

4. Were any of the species that you identified outside of their normal range? If so, which ones?

_____

_____

E A bird outside its species' normal range is called a *vagrant*. Do some research on bird vagrancy.

5. Write a paragraph summarizing some of the causes of vagrancy in birds.

_____

_____

_____

_____

_____

_____

_____

F Compare the species that you identified with those your classmates identified.

6. Which species was identified by the most people?

_____

G For each of those species for which two or more individuals were observed by the class as a whole, add the total number of individuals and record the common name and the number—the value for *n*—in the first two columns of Table 2.

7. Which species had the highest total number of individuals counted by your class?

_____

H Create a bar graph in Graphing Area A using the results that you obtained in Step G.

ANALYSIS

Do you remember Simpson's Index from Lab 4A? It uses the formula

$$D = \frac{\Sigma n(n-1)}{N(N-1)},$$

where D is the index value, the Greek letter sigma (Σ) means "sum of," n is the number of individuals of any one species in the sample, and N is the total number of organisms in the sample.

I Calculate the value of Simpson's Index for the data that you recorded in Table 2.

(1) Beginning with the first species listed in your sample, record the value for $n-1$ and $n(n-1)$ in Table 2. Repeat this step for each species in your sample.

(2) Add up all the values for n; this sum is N. Record the value for N in Table 2.

(3) Add up all the values for $n(n-1)$ calculated in Step (1). This is $\Sigma n(n-1)$. Record this value in Table 2.

(4) Multiply the total number of individuals of all species in your sample (N) times that number minus one ($N-1$). This is $N(N-1)$. Record this value in Table 2.

(5) Divide the value for $\Sigma n(n-1)$ calculated in Step (3) by the calculated value for $N(N-1)$ from Step (4) to obtain Simpson's Index (D) for your sample. Record this value in Table 2.

(6) Simpson's Diversity Index, the actual number that you need to assess the diversity of your sample, is slightly different from Simpson's Index. To obtain the value of Simpson's Diversity Index for your sample, subtract the value that you calculated in Step (5) from 1 (i.e., $1-D$).

8. What is the value of Simpson's Diversity Index for your sample?

_____

You may remember from Lab 4A that a low value for Simpson's Diversity Index indicates a low biodiversity.

9. What does the Simpson's Diversity Index value that you calculated for your sample tell you about the avian biodiversity of your sample?

_____

GOING FURTHER

A study published in Royal Society Open Science in 2016[1] considered populations of English wrens using data collected by volunteer birders. The scientists who analyzed that data found that more wrens lived in southern Britain than in northern Britain and that the wrens in the north tended to be larger than their cousins in the south.

10. How was the work of volunteers valuable to the scientists involved with this study?

_____

_____

11. The scientists who conducted this study concluded that the size difference was a result of adaptation to climate differences. How does this line up with a biblical worldview?

_____

_____

_____

_____

[1] Morrison, Catriona A., Robinson, Robert A., Pearce-Higgins, James W., "Winter Wren Populations Show Adaptation to Local Climate," Royal Open Science, Vol. 3, No. 6 (June 2016).

TABLE 1 Observed Bird Species

	COMMON NAME	SCIENTIFIC NAME	HABITAT	LOCATION	NUMBER
1					
2					
3					
4					
5					
6					
7					
8					
9					
10					

TABLE 2 Bird Count Data

COMMON NAME	NUMBER (n)	$n - 1$	$n(n - 1)$
	$N = $ _____	$N(N - 1) = $ _____	$\Sigma n(n - 1) = $ _____
			$D = $ _____

GRAPHING AREA A

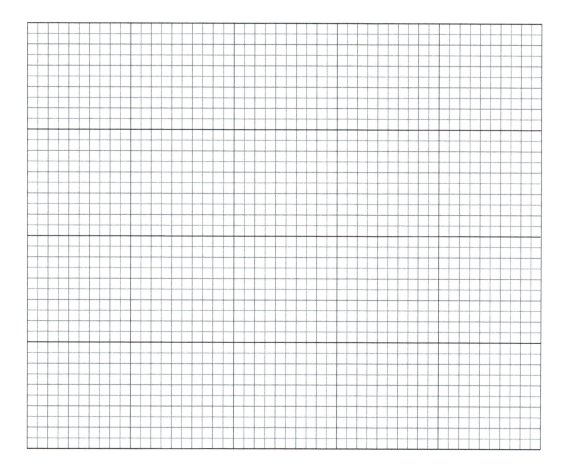

19B LAB

Warming Up to Research

Doing Preliminary Research

Scientists are curious people—they wonder about things that other people might not ever think about. Do dogs dream in color? Can an owl survive on a vegetarian diet? Do bears get grumpy if they are awakened in the middle of hibernation? These are the kinds of questions that motivate scientists to keep acquiring ever more knowledge about how God's creation works.

But how does a scientist know for certain that a question has never been answered before, perhaps not even investigated? Funding for scientific research is often limited, and the people who finance research are probably not interested in funding studies whose questions have already been asked and answered. And what if a question *has* been asked and investigated but not answered? How can a scientist know what kinds of investigative work have already been done? After all, there's no need to start at square one if a scientist can build on someone else's foundation.

The answer to these questions is research. Preliminary research is one of the early steps in the process of scientific inquiry. Good preliminary research helps scientists figure out where to focus their efforts and how to proceed with an investigation. In this activity you'll have a chance to practice this very important scientific skill.

How do I find research that has already been done on a scientific question?

Equipment
computer with internet connection

QUESTION
- How do scientists narrow down the focus of their research?

PROCEDURE

A Think of an endothermic animal that interests you—be specific. Then think of some questions that you could ask about that animal. Do a preliminary internet search on current research to get an idea of what some scientists are presently investigating. Narrow down your list of questions to the one that interests you the most.

1. What animal will you be researching?

_____

2. What question will you research?

_____

_____

B Now do an internet search for information related to your specific question.

3. Choose three articles, then cite each one and write a brief summary of the article's topic and findings.

   a. Article: _____

   Summary:

   b. Article: _____

   Summary:

   c. Article: _____

   Summary:

4. Propose a hypothesis on the basis of your research that, if tested, could expand what is already known about the subject under investigation.

   _____

   _____

5. On the basis of what you have learned from your research, suggest another avenue of investigation for any future scientist who may be interested in the subject.

   _____

   _____

   _____

   _____

20A LAB

Chill Out!

Inquiring Into the Skin's Ability to Maintain Homeostasis

Have you ever been caught outside when the temperature suddenly fell? Perhaps you've had the unhappy experience of forgetting to bring a coat to a late-autumn football game. As the chill sets in, your body's heat conservation mechanisms activate in an attempt to maintain homeostasis. Your skin is the first line of defense. The blood vessels that supply your skin constrict, reducing the flow of blood to your skin's surface. Your fingers and toes begin to feel numb, and your skin is cold to the touch.

These early effects of cold on the human body, if tended to, are not permanent of course. Wrapping oneself in a thick blanket and enjoying a cup of hot cocoa will quickly put things right. The constricted blood vessels reopen, and soon your fingers and toes feel normal again. It doesn't happen instantly though. It takes a bit of time. But does every part of your body require the same amount of time to recover its normal temperature after being chilled? That's the question you'll explore in this lab activity.

Does chilled skin on different parts of the body recover temperature at the same rate?

Equipment

to be determined by your investigation

QUESTION

- Are all parts of the skin equally effective at regulating temperature?

PROCEDURE

Plan and Write Scientific Questions

A Think about how different parts of your body respond after being chilled, then brainstorm with your group to come up with a list of possible questions to investigate. Consider which ones interest you the most. You'll need to take into account whether you have the necessary equipment on hand (or can easily obtain it) and enough time to adequately test your possible hypotheses. Narrow down your list of questions to the one that you will investigate.

1. Which areas of skin will you be comparing?

_____

2. Write a hypothesis about how one area of skin's temperature recovery will compare with another's.

_____

_____

Designing and Conducting Scientific Investigations

Next, design an experiment to test your hypothesis. Get approval for your experimental design from your teacher before beginning your experiment.

3. Briefly describe your experimental setup.

_____

_____

_____

4. What will be your control group?

_____

5. What will be your experimental group?

_____

6. What data will you be collecting?

_____

_____

_____

7. How will you analyze your data?

_____

_____

_____

Set up and run your experiment. Make sure to run several trials to solidify support for your conclusions.

Developing Models

Once you have collected your data, you will need to analyze it by using mathematical tools, such as finding and comparing averages or creating graphs.

Scientific Argumentation

In Lab 15B you learned about scientific writing. Recall that a scientific paper has distinct sections, including an abstract, an introduction, a discussion of materials and methods, the results, a discussion of the meaning of the results (including suggestions for further research), and a list of references. Your teacher will instruct you on whether to report your findings in a formal lab report (a type of scientific paper), an oral report, or a slide presentation.

20B LAB

Are You Aware?

A Case Study in Advocacy

How should a Christian view health awareness issues?

Equipment
none

LYMPHATIC SYSTEM DISORDERS

As you learned in Chapter 20, lymph is a fluid that fills the spaces in between cells. It is regularly collected via the lymphatic system and returned to the bloodstream. But sometimes this collection system doesn't work the way it's supposed to. In such instances lymph accumulates in one or more parts of the lymphatic system, causing edema (swelling). This condition is known as *lymphedema*. As the disease progresses, the affected part of the body becomes greatly enlarged, with dark and thickened skin. In severe cases limbs become deformed. The affected person may no longer be able to wear normal-fitting clothes or shoes. Lymphedema may hinder the person's ability at work or in enjoyment of favorite activities and hobbies.

No one knows what the exact cause of lymphedema is. It can be inherited, in which case it is known as primary lymphedema. Noninherited, or secondary, lymphedema, is most often seen in cancer patients, especially those who have had radiation therapy or lymph nodes removed as part of their treatment. It is especially common in breast cancer patients.

Lymphedema is not the only lymphatic system disorder, nor is it the most serious. Diseases of the lymphatic system range from swelling of the lymph nodes in response to an infection to serious cancers such as Hodgkin's lymphoma. Do an internet search on lymphatic system disorders. Choose one disorder to research, and then answer the following questions.

QUESTIONS

• What is lymphedema?

• What is advocacy?

• How can Christians decide where help is needed most?

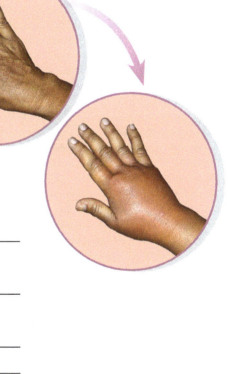

1. What lymphatic system disorder did you research?

_____

2. What part of the lymphatic system is affected by this disorder?

_____

3. What are the symptoms of this disorder?

_____

_____

_____

4. What treatments are used for this disorder, if any?

_____

_____

_____

5. What are some current areas of research with regard to this disorder?

_____

_____

_____

A LOOK AT ADVOCACY

No one knows exactly how many people are affected by lymphedema, but many experts believe that the disease is underreported. Some estimates put the number of cases at as many as 250 million worldwide. Despite its prevalence, there's a good chance that you've never heard of this disease. That shouldn't be too surprising. After all, there is a huge number of diseases and disorders out there!

And therein lies a problem. There are thousands of different kinds of diseases and disorders, and hundreds of millions of people are affected by them, but the resources for research into finding cures are not unlimited. How can the victims of these diseases and the people who treat them obtain the resources they need to combat the disease? One way to do this is through advocacy.

Advocacy is a concerted effort to promote support for a particular cause. It's likely that you've seen a pink ribbon sticker on a car or that your favorite sports team has a day during their season when they wear pink as part of their uniform. These symbols are part of the effort to raise awareness of and financial support for breast cancer research and treatment. This effort has recruited many thousands of volunteers and raised many millions of dollars. Yet the issue of advocacy raises some interesting questions. Think about the following questions and be prepared to share your responses with your classmates.

▲ _____
A blue ribbon is the symbol for lymphedema awareness.

6. Why is it worthwhile to spend money on finding cures for diseases?

_____

_____

_____

_____

_____

7. As of 2015 there were 212 health awareness days, weeks, or months recognized by the United States Federal Government, such as National Folic Acid Awareness Week. Do you think that this is an effective strategy for raising health awareness? Explain.

_____

_____

_____

8. Why do you think Congress has decreed so many health awareness dates?

_____

_____

_____

9. What might be some of the weaknesses in an awareness campaign for raising funds and support for treating disease?

_____

_____

_____

_____

10. Read Mark 1:29–39. What does this passage suggest about how Christians should balance health awareness and other Christian activity?

_____

_____

_____

_____

11. What is the best way for Christians to wisely use their money to promote health-based charity?

_____

_____

_____

_____

GOING FURTHER

In 2016 the US Senate officially recognized World Lymphedema Day, which is observed in early March of each year. With so many health awareness days being set aside each year, there are probably several that you have never heard of before. Do an internet search on national health observances and find a health awareness day that interests you but with which you are unfamiliar. Find out what health issue the day targets and learn a little about it. Discover what organizations promote the day and what sorts of activities they organize in order to increase public awareness. Summarize your findings in a paragraph or two and be prepared to share your findings with your classmates.

21A LAB

Dry Bones

Exploring the Skeletal System

People who have experienced one know that a broken bone is no fun. Although broken bones are not necessarily uncommon, the fact that they do not occur more frequently is a testament to God's marvelous design. It takes a lot of force in the right direction to break a bone. The bones, along with the ligaments, joints, and cartilage, provide the support that the body needs. Today you will have an opportunity to study the skeleton at both the microscopic and the macroscopic levels.

How do the parts of the skeletal system work together to support the human body?

Equipment

microscope

preserved slides of dry ground bone, c.s.

model of a human skeleton

skeleton diagrams

PROCEDURE

Microstructures

A Observe the preserved slide of dry ground bone, c.s., and draw a diagram of a complete Haversian system in Drawing Area A. Label your diagram. Include the Haversian canal, lamellae, lacunae, canals, and any other structures observed in the preserved specimen.

Macrostructures

B Memorize the bone names and markings indicated on page 478 of your textbook.

C Locate and label these bones on the models of the skeletons on page 225 of this lab activity.

D Memorize the types of joints on page 479 of your textbook. Fill in the missing information in Table 1.

1. How does the structure of the bones show God's care for His creatures?

_____

_____

_____

QUESTIONS

- What is the microstructure of a typical bone?

- What are the bones of the skeletal system?

- What are the sesamoid bones?

GOING FURTHER

Do an internet search on sesamoid bones, and then answer the following questions.

2. What are sesamoid bones?

_____

_____

3. How do sesamoid bones develop?

_____

_____

4. List some examples of sesamoid bones.

_____

_____

DRAWING AREA A

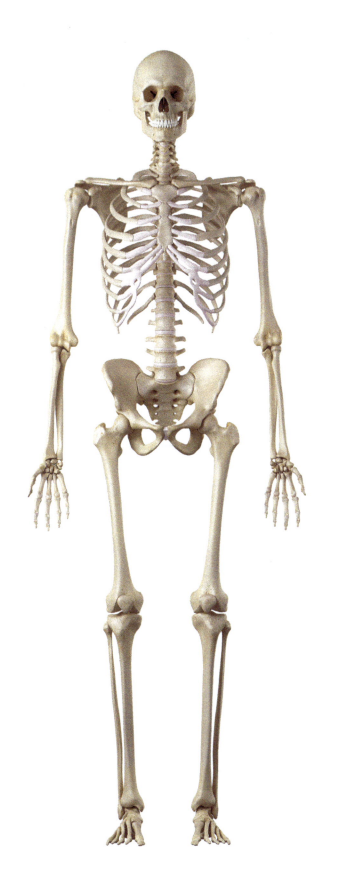
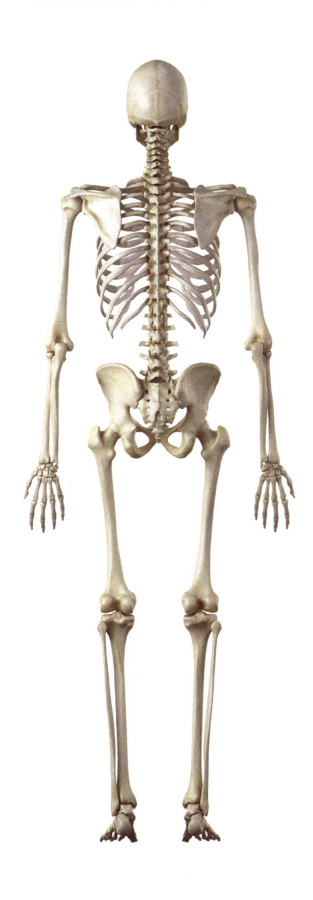

TABLE 1 Joints of the Human Body

CATEGORY OF JOINT	TYPE OF JOINT	DESCRIPTION OF MOVEMENT	EXAMPLES
Movable		rotation and free movement in all directions	
	hinge		
			the two vertebrae at the base of the skull
	gliding		
Slightly Movable	slightly movable		
Immovable		no movement (because two bones have grown together)	

21B LAB

I'm So Tired!

Investigating Muscular Function

We rely on muscles to allow us to move, but our muscles can take only so much. Sooner or later, a muscle used too much can move no more and must rest. This is called *muscle fatigue*. Although the mechanisms of muscle fatigue are not completely understood, it is known that people such as athletes, who use their muscles often, are able to use their muscles longer before getting muscle fatigue.

In this lab activity you will be experimenting with muscle fatigue and some of the factors that can improve or hurt muscle performance.

What factors affect muscular function?

Equipment

plastic jug, 1 gal, filled with water
stopwatch
hand gripper

dishpan
cold water, approximately 10 °C

PROCEDURE

Muscle Fatigue in the Dominant and Nondominant Arms

A Extend your dominant arm parallel to the ground while holding the gallon jug of water.

B Have your lab partner time how long you can keep your arm in its parallel position.

C As soon as your arm drops out of its parallel position, have your lab partner note the time and record it in Table 1.

D Repeat Steps A through C with your nondominant arm.

E Repeat Steps A through D an additional four times.

F Using a graphing calculator or spreadsheet, calculate and record the average and standard deviation of your data for each arm in Table 1.

G Graph your results from Step F on Number Line 1. Since the time that a person can hold up a jug of water can vary quite a bit, you will create your own scale for the number line.

1. Considering the data in your number line, can you say that one arm was stronger than the other? Explain.

_____

_____

_____

> **QUESTIONS**
> - What is muscle fatigue?
> - How does muscle fatigue affect muscular function?
> - How does temperature affect muscular function?

2. If one of your arms was stronger than the other, was your dominant or non-dominant arm stronger?

_____

3. Suggest a possible reason why you observed the results that you reported in Questions 1 and 2.

_____

_____

_____

_____

Muscle Fatigue and Muscle Strength

H Count the number of times that you can squeeze a hand gripper in 30 seconds and record the number in Table 2.

I Using the same arm, repeat the experiment an additional nine times with as little time as possible between each trial.

J Graph the results with a scatterplot in Graphing Area A. Create a line of best fit for your data.

4. Did your ability to squeeze the hand gripper change over time? If so, how did it change?

_____

_____

_____

5. Do some research on exercising. How can a person develop the ability to work a muscle for a longer period of time without experiencing muscle fatigue?

_____

_____

_____

Temperature and Muscle Fatigue

K After your hand has rested, squeeze the hand gripper as rapidly as possible for 30 seconds and record the number of squeezes in the _Warm Arm_ column of Table 3.

L Repeat Step K an additional four times.

M After your hand has rested, submerge your forearm in cold water for 1 minute. Then, using the same arm, squeeze the hand gripper as rapidly as possible for 30 seconds and record the number of squeezes in the _Cold Arm_ column of Table 3.

N Repeat Step M an additional four times.

O Find the average for both the warm arm and the cold arm. Record these averages in Table 3.

P Graph the results using a bar graph in Graphing Area B.

6. Why do you think you were instructed to chill your forearm instead of your hand?

_____

_____

7. On the basis of your data and your graph, do you think that there is any meaningful difference between the time it took for muscle fatigue to set in for a warm muscle compared with a cold muscle?

_____

_____

8. On the basis of your data, suggest a relationship between muscle temperature and fatigue.

_____

_____

_____

_____

9. Stretching is important before exercising. Do some research on why it is important to perform a warmup routine before stretching. Write two or three sentences on what you find.

_____

_____

_____

_____

10. What are some fields of work that require knowledge of muscle function and muscle fatigue?

_____

_____

11. Choose one of the fields from your answer to Question 10 and write a paragraph on how a person in that field can use knowledge about muscle function and muscle fatigue to show love to other people.

_____

_____

_____

_____

_____

TABLE 1 Muscle Fatigue in the Dominant and Nondominant Arms

TRIAL #	DOMINANT ARM (time in seconds)	NONDOMINANT ARM (time in seconds)
1		
2		
3		
4		
5		
Average		
Standard Deviation		

NUMBER LINE 1

TABLE 2 Muscle Fatigue and Muscle Strength

TRIAL #	NUMBER OF SQUEEZES IN 30 s
1	
2	
3	
4	
5	
6	
7	
8	
9	
10	

TABLE 3 Temperature and Muscle Fatigue

TRIAL #	WARM ARM (number of squeezes)	COLD ARM (number of squeezes)
1		
2		
3		
4		
5		
Average		

GRAPHING AREA A

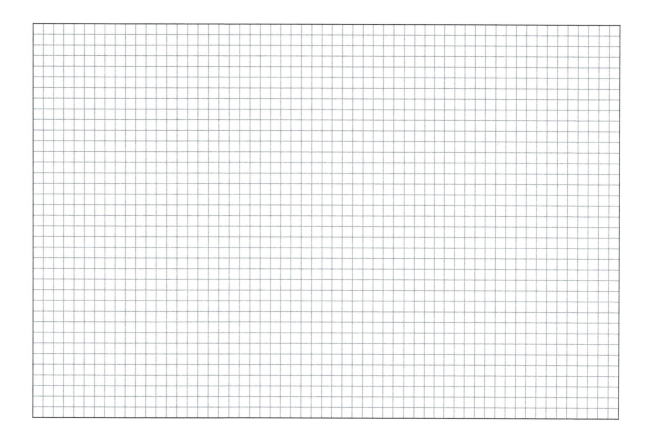

GRAPHING AREA B

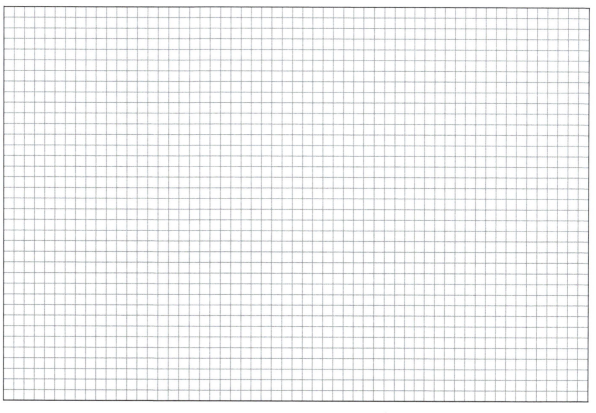

22A LAB

Relax and Take a Deep Breath

Exploring the Human Respiratory System

We've all been told to relax and take a deep breath at one time or another, usually right before we have to do something unpleasant or stressful—like share a lab report with our classmates! But exactly how deep a breath can a person take? And does your friend's deep breath take in the same amount of air as your deep breath? Is there even a way to tell? It turns out there is, and looking at breath volume is one of the things that you'll do during this activity.

We don't give too much thought to breathing. It's an involuntary activity, one that doesn't require conscious thought to control. That's a good thing—otherwise it might be difficult to sleep at night! Your brain's respiratory centers watch over every breath you take—all 600,000,000 or more over the course of an average lifetime. While those respiratory centers go about their business, let's take a look at the respiratory system.

> **How much air can an average person inhale in one breath?**
>
>

Equipment

lung volume bag
mouthpieces (3)
rubber band

stethoscope
isopropyl (rubbing) alcohol
tissues

paper towels

> **QUESTIONS**
> - What is an average lung capacity?
> - What are the structures of the respiratory system?

PROCEDURE

Measuring Tidal Volume

A Prepare the lung volume bag. Select one person from your group to be tested and another to act as assistant.

> » Insert the mouthpiece partway into the open end of the lung volume bag and secure it with a rubber band.

> » Have the assistant sit down.

> » Slide the bag slowly across the assistant's knee while the assistant presses the bag with a paper towel. This will remove all air from the lung volume bag.

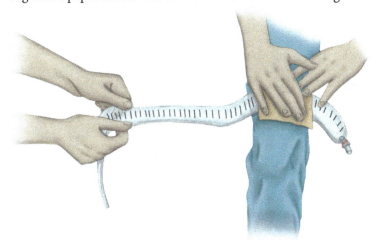

B The person being tested should breathe in a normal breath, pinch the nose, put the lung volume bag mouthpiece in the mouth, and breathe out a normal breath. This normal breath is known as the *tidal volume* of the lungs.

C The assistant should take the bag and hold it closed while sliding it across the knee and pressing it with a paper towel in order to force all the air to the closed end. Note how many liters of air are in the bag and record the data in the Tidal Volume section of Table 1.

D Empty the lung volume bag, using the procedure described at the end of Step A.

E Repeat Steps B through D for two more trials.

F Using new mouthpieces, repeat Steps B through E for the other two students. Students should hold on to their mouthpieces.

Measuring Expiratory Reserve Volume

G Prepare the lung volume bag using the first student's mouthpiece.

H The first student should next breathe in a normal breath, breathe out a normal breath, pinch the nose, put the lung volume bag mouthpiece in the mouth, and breathe out as much as possible. This extra bit of breath is known as the *expiratory reserve volume*.

I Using the procedure described in Step B, measure the air and record the volume in the Expiratory Reserve Volume section of Table 1.

J Repeat Steps G–I for two more trials.

K Repeat Steps G–J for the other two students. Make sure that students are using their own mouthpieces for their trials.

Measuring Vital Capacity

L Prepare the lung volume bag using the first student's mouthpiece.

M The person being tested should now breathe in as much as possible, pinch the nose, put the lung volume bag mouthpiece in the mouth, and breathe out as much as possible. This maximum-sized breath measures the lungs' *vital capacity*.

N Using the procedure described in Step B, measure the air and record the volume in the Vital Capacity section of Table 1.

O Repeat Steps L–N for two more trials.

P Repeat Steps L–O for the other two students.

Computing Average Lung Capacities

Q Determine the average for each of the volumes.

1. What is the average tidal volume for the students tested?

_____

2. What is the average expiratory reserve volume?

_____

3. What is the average vital capacity?

_____

The additional air that can be forcibly inhaled after the inspiration of a normal tidal volume is known as the *inspiratory reserve volume*.

4. Using the answers to Questions 1–3, determine the average value for the inspiratory reserve volume for the students tested.

_____

5. Assuming a residual volume—the amount of air left in the lungs after a person has forced as much out as possible—of 1000 mL, what would be the average *total lung capacity* of the people tested?

_____

6. According to the information on page 496 of your textbook, are these lung volume results considered average? If not, what factors might account for the difference?

_____

_____

_____

_____

Listening to Your Lungs

Air rushing into and out of healthy respiratory structures makes various sounds. Many respiratory problems cause abnormal sounds that a physician can hear using a stethoscope. Hopefully, you will not hear abnormal respiratory sounds when listening to your lab partner's breathing! But it can be interesting to listen to normal sounds of the respiratory system.

R Using a tissue and alcohol, clean the earpieces of the stethoscope.

S Place the stethoscope in your ears, allowing the tubes to hang freely and being careful not to hit the diaphragm on hard objects. (The noise can be very loud.)

T Place the diaphragm of the stethoscope on your lab partner's back and press lightly as he or she breathes normally.

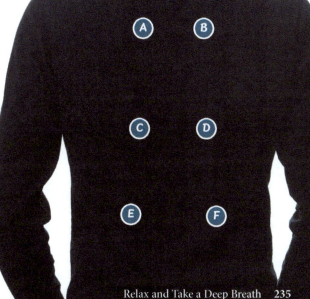

7. Listen to a normal breath or two in Areas A–F as labeled on page 235. Describe the sounds you hear.

_____

_____

_____

8. Are the sounds you hear different in different areas? Explain.

_____

_____

_____

9. Listen to a deep breath in the same areas. Do you hear a difference? If so, describe the difference and tell what may account for it.

_____

_____

_____

10. Listen to Areas A and E while your partner coughs. Describe what you hear.

_____

_____

_____

11. Listen to Areas A and E while your partner talks. Describe what you hear.

_____

_____

_____

U Now listen to your lab partner's breathing by placing the stethoscope in Area G. Listen to your lab partner's voice in the same area.

12. Describe what you hear.

_____

_____

_____

Structures of the Respiratory System

Label the diagram below as completely as possible. Under each label indicate the function of that structure (enclose this information in parentheses).

13. In the Greek New Testament the word *pneuma* is used for both the Holy Spirit and for breath. How are the work of the wind (breath) and the Holy Spirit compared in John 3:5–8?

_____

_____

_____

_____

_____

_____

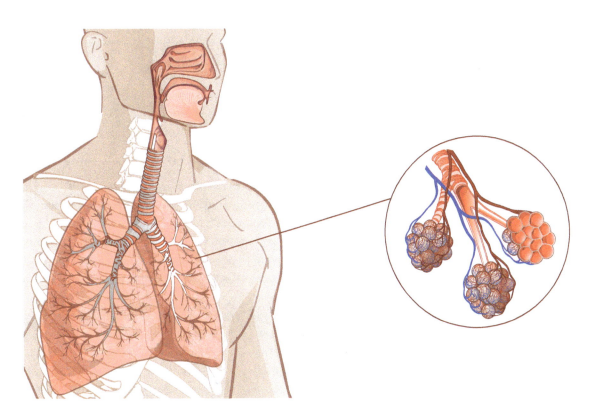

TABLE 1 Lung Volume

PERSON A	PERSON B	PERSON C	AVERAGE
Tidal Volume			
1. _____	1. _____	1. _____	_____ mL
2. _____	2. _____	2. _____	or
3. _____	3. _____	3. _____	_____ L
Expiratory Reserve Volume			
1. _____	1. _____	1. _____	_____ mL
2. _____	2. _____	2. _____	or
3. _____	3. _____	3. _____	_____ L
Vital Capacity			
1. _____	1. _____	1. _____	_____ mL
2. _____	2. _____	2. _____	or
3. _____	3. _____	3. _____	_____ L

22B LAB

Feeling the Pressure

Investigating Blood Pressure and Hypertension

Shelana was a 36-year-old mother of four. She had spent a busy day planning for her daughter's tenth birthday, in spite of having had a terrible headache for the previous two weeks. Shelana suffered from migraines, so she hadn't thought too much about the headache, but that night over dinner she mentioned her intention to see her doctor the next day.

As Shelana was drifting off to sleep later that night, she felt what she described as a "weird" sensation. She tried to tell her husband about it, but her speech was slurred and incoherent. Her husband rushed her to an emergency room. After a night in the hospital and an MRA (a type of scan that provides pictures of blood vessels), a diagnosis was made. Shelana had suffered two ischemic strokes. Two weeks in intensive care and many months of physical therapy followed. Even two-and-a-half years later she still did not have full use of the right side of her body.

A stroke occurs when blood flow to the brain is reduced, either due to a clot (an ischemic stroke) or bleeding (a hemorrhagic stroke). In 2021 strokes were the fifth-leading cause of death in the United States. We sometimes think of a stroke as something that happens to older people, but as Shelana's case shows, a stroke can happen to anyone.

A *risk factor* is anything in a person's life that increases the risk of having a particular medical condition. The greatest risk factor for stroke is high blood pressure, or hypertension. Blood pressure measurement shows how much pressure is exerted by blood on the walls of a person's blood vessels. Let's take a closer look at this vital topic.

How can I lower my risk of having a stroke?

Equipment
digital blood pressure cuff

QUESTIONS
- What is blood pressure?
- What is hypertension?
- What are some health risks associated with high blood pressure?

PROCEDURE

Measuring Blood Pressure

To obtain a blood pressure measurement, follow the instructions provided with the digital blood pressure cuff given to you by your teacher. Blood pressure measurements are generally taken in a relaxed, seated position. At the conclusion of the test, the digital blood pressure cuff will display two numbers. The larger of the two numbers is the *systolic pressure*, the pressure within your arteries when your heart is contracting. This pressure is measured in millimeters of mercury (mmHg).

A Obtain a blood pressure measurement.

1. What is your systolic pressure?

_____

The smaller number is your *diastolic pressure*, the pressure within your arteries when your heart is between beats.

2. What is your diastolic pressure?

_____

B Obtain two additional blood pressure measurements.

3. Are the values for the second and third measurements the same as for the first measurement?

_____

4. Suggest a way in which to report a more reliable blood pressure measurement.

_____

_____

_____

Blood pressure measurements are categorized according to their indications for hypertension. The following chart reflects categories defined by the American Heart Association.

TABLE 1 Blood Pressure

BLOOD PRESSURE CATEGORY	SYSTOLIC (mmHg)		DIASTOLIC (mmHg)
Normal	less than 120	and	less than 80
Elevated	120–129	and	less than 80
High Blood Pressure Stage 1 Hypertension	130–139	or	80–89
High Blood Pressure Stage 2 Hypertension	140–180	or	90–120
Hypertension Crisis (emergency care needed)	higher than 180	and/or	higher than 120

5. Using the information in Table 1, indicate which blood pressure category you are in.

_____

6. What are some reasons why that category might not accurately describe your current blood pressure health?

_____

_____

_____

_____

_____

_____

Assessing Risk for High Blood Pressure

As we saw in the opener, having high blood pressure increases a person's chances for having a stroke. Do an internet search for high blood pressure risk factors, and then answer the following questions.

7. What are some risk factors for high blood pressure?

_____

_____

_____

_____

_____

8. Can the cause of a person's high blood pressure always be determined? Explain.

_____

_____

9. Besides stroke, what are some other health risks associated with high blood pressure?

_____

_____

_____

A severe stroke victim faces the possibility of a very long recovery period, perhaps including the need to relearn fine motor skills.

10. What are some lifestyle choices that you can make right now to reduce your risk for high blood pressure?

_____

_____

_____

_____

11. Why should Christians be concerned about maintaining healthy blood pressure and reducing risks?

_____

_____

_____

_____

23A LAB

Calorimetry in a Can

Measuring the Energy in Food

It's break time at your school and time for a snack. But among the seemingly infinite variety of snack foods available today, which one will provide you with the biggest energy boost? Let's find out!

Which snack food delivers the most energy?

Equipment

laboratory balance (accurate to 0.01 g)

evaporating dish, 100 mL

ring stand

graduated cylinder, 50 mL

thermometer or temperature probe

iron ring

cork stopper

large straight pin

glass stirring rod

empty soda can with attached tab

grill lighter

tongs or mitts

various snack foods with nutrition facts labels

laboratory apron

goggles

QUESTIONS

- How is a soda can calorimeter used to determine the energy content of food?

- Which snack foods contain the most energy?

- How accurate is a soda can calorimeter?

PROCEDURE

Calorimetry is the technique that we'll use to determine the amount of energy in a sample of your favorite snack. A calorimeter burns a small sample of food to heat a known quantity of water. The resulting temperature change of the water is then used to calculate the energy released by the burning sample.

A Create a food holder by pushing the straight pin through the cork stopper as shown on the right. Push the pin through the center of the cork, if possible; otherwise, push it through from one side.

B Secure a small piece of the first snack food item to be tested on the point of the pin, then place the food holder on the evaporating dish.

C Use the balance to determine the mass of the evaporating dish, food holder, and food sample. Record this as the initial mass in Table 1.

D Place the evaporating dish on the base of the ring stand.

E Use the graduated cylinder to add 50.0 mL of water to the soda can.

F Measure the temperature of the water to the nearest 0.1 °C and record the result in Table 1.

G Bend the tab of the soda can upward so that the stirring rod will easily pass through it. Slide the rod through the tab and suspend the can from the ring of the ring stand, leaving about 2 cm between the bottom of the can and the top of the food sample.

H Use the grill lighter to ignite the food sample. As soon as the sample finishes burning, measure the temperature of the water in the soda can, again to the nearest 0.1 °C. Record this final temperature value in Table 1.

I With a set of tongs or mitts, move the evaporating dish and food sample to the balance and determine the final dish and food mass. Record this result in Table 1.

J After making sure that the evaporating dish has cooled, clean the dish and food holder, then repeat Steps B–I for the remaining two snack food items.

ANALYSIS

Using the data that you collected, you can calculate the amount of heat energy that was released when the food was burned.

K For each snack food test, subtract the starting temperature of water from its final temperature and record the temperature change (Δt) in Table 2.

L For each snack food item, calculate the mass of food consumed during the test by subtracting the final mass of the dish, holder, and sample from its starting mass. Record the mass in Table 2.

The formula for heat energy is

$$Q = c_{sp}m\Delta t,$$

where Q is the heat energy in calories absorbed by the water, c_{sp} is the specific heat of water (1.0 cal/g °C), m is the mass of the water, and Δt is the change in temperature of the water.

M Calculate the heat energy absorbed by the water from each snack food sample. Nutrition fact panels typically use the Calorie (note the capital C) as a standard unit of heat energy. A Calorie is 1000 calories, so divide each value for Q by 1000 and record the results in Table 2.

N Find each snack food's energy per unit of mass by dividing the value for Q from Step M by that snack food's change in mass. Record these values in Table 2.

1. Did any of the snack foods have a substantially higher energy content per gram than the others? If so, which one?

_____

O In Table 2 record the reported energy per unit of mass for each snack food in Calories per gram. You will probably need to calculate these values on the basis of the serving size and calories per serving found in each snack's nutrition facts label.

P Calculate the percent error of your calorimetry results by using the formula below.

$$\% \ error = \left(\frac{|value_{estimated} - value_{actual}|}{value_{actual}} \right) 100\%$$

The estimated value is the result obtained from your calorimetry test, and the actual value is the value obtained from the nutrition facts label on the snack food packaging. Record your results in Table 2.

GOING FURTHER

2. According to your data, is a soda can calorimeter reliable for determining the energy content of snack foods?

_____

_____

_____

_____

_____

3. According to your data, is a soda can calorimeter useful for comparing the energy contents of different snack foods?

_____

_____

_____

_____

_____

4. How do you suppose snack food producers determine the reported energy contents of their products?

_____

_____

_____

5. How can knowing the energy content of a snack food help you make healthier living choices?

_____

_____

_____

TABLE 1 Calorimetry Data

Snack Food	Initial Mass (dish, food holder, and sample) (g)	Initial Water Temperature (°C)	Final Water Temperature (°C)	Final Mass (g)

TABLE 2 Analysis

Snack Food	Temperature Change (Δt) (°C)	Mass of Food Consumed (g)	Energy Released (Q) (Cal)	Energy per Unit of Mass, Experimental (Cal/g)	Energy per Unit of Mass, Reported (Cal/g)	% error

23B LAB

What a Waste!

Modeling Dialysis

Do you remember your study of osmosis in Lab 5B? Osmosis occurs when the water from a solution passes through a semi-permeable membrane but the solute cannot. But what happens when some of the solutes can also pass through the membrane? This process is called *dialysis*, and it is the basis for the cleansing action of the kidneys.

You will be working with a mixed solutes solution containing the protein albumin, the carbohydrates glucose and starch, and the salt sodium chloride.

How do the kidneys use dialysis to remove wastes from the blood?

Equipment

multimeter	disposable pipettes	iodine solution
beakers, 250 mL (3)	graduated cylinder, 100 mL	paper towels
test tubes (9)	hot water bath	laboratory apron
test tube rack	mixed solutes solution	nitrile gloves
test tube holder	distilled water	goggles
dialysis tubing strip	biuret reagent	
clamps (2)	Benedict's solution	

PROCEDURE

QUESTIONS

- What is dialysis?

- Which solutes will pass through a semipermeable membrane?

- What factors determine which molecules can pass through a semipermeable membrane?

Dialysis Setup

A Label three 250 mL beakers A, B, and C.

B Label the nine test tubes A1, A2, A3, B1, B2, B3, C1, C2, and C3 and place them in the test tube rack.

C Obtain a 12 cm strip of dialysis tubing that has been soaked in water. Close one end of the strip with a clamp to form a bag with an open end.

D Using a pipette, place 10 mL of the mixed solutes solution in the dialysis bag.

E Clamp the other end of the tubing, leaving about 2 cm of air in the bag. Rinse the bag with distilled water and blot dry with a paper towel.

F Place the bag in Beaker A with 200 mL of water. Allow the bag to soak in the water overnight.

1. Form a hypothesis about which solutes will pass through the membrane.

_____

_____

Positive and Negative Tests

While the dialysis bag is soaking, you can examine what positive and negative tests look like for each of the solutes in the mixed solutes solution.

G Pour 100 mL of water into Beaker B.

H Pour 10 mL of water into each of Tubes B1, B2, and B3.

I Use the graduated cylinder to pour 100 mL of the mixed solutes solution into Beaker C.

J Use a pipette to transfer 10 mL of the mixed solutes solution into each of Tubes C1, C2, and C3.

Biuret Reagent for Albumin

Biuret reagent is a liquid that changes from blue to pink or purple in the presence of proteins such as albumin.

K Use a pipette to add 2–5 drops of biuret reagent to Tubes B1 and C1.

L Check whether the solutions test positive or negative for albumin and record your results in Table 1. Enter a "+" for a positive test and a "–" for a negative test.

Benedict's Solution for Glucose

Benedict's solution is a blue liquid. When mixed with monosaccharides such as glucose and heated, Benedict's solution changes to an orangish red.

M Use a pipette to add 2 mL of Benedict's solution to Tubes B2 and C2.

N Place both test tubes in a hot water bath for 10 minutes.

O Check whether the solutions test positive or negative for glucose and record your results in Table 1.

Iodine Solution for Starch

Iodine solution is a brown liquid that changes to blue or purple in the presence of starch.

P Use a pipette to add 2–5 drops of iodine solution to Tubes B3 and C3.

Q Check whether the solutions test positive or negative for starch and record your results in Table 1.

Electrolyte Test

Although pure water is a poor conductor of electricity, dissolved electrolytes, such as sodium and chloride ions, which result when NaCl is dissolved in water, make it a good conductor. You can take advantage of this fact to determine whether the sodium and chloride ions passed through the dialysis tubing into Beaker A. You will use a multimeter set to measure resistance to test whether the water will conduct an electric current. If the water has electrolytes in it, the multimeter will indicate a certain number of ohms (Ω) of resistance. If the water contains no electrolytes, the multimeter will indicate either 0 Ω or infinite Ω (depending on the multimeter).

R Use the multimeter to check whether the contents of Beakers B and C test positive or negative for the presence of electrolytes, and record your results in Table 1. Be sure to rinse the probes between tests.

Testing Dialysis

After your dialysis bag has soaked overnight, it's time to test the water from Beaker A.

S Remove the dialysis bag and set it aside.

T Use a clean pipette to place 10 mL of water from Beaker A into each of Tubes A1, A2, and A3.

U Following the steps from the previous procedures, check Tube A1 for albumin using biuret reagent, Tube A2 for glucose using Benedict's solution, and Tube A3 for starch using iodine solution. Record the results in Table 1.

V Check Beaker A for sodium and chloride ions using the multimeter. Record the results in Table 1.

2. Which solutes passed through the dialysis membrane?

_____

3. Why do you think that some solutes passed through the membrane while others did not?

_____

_____

_____

_____

4. How do the kidneys use dialysis to cleanse the blood of wastes? You may need to reference your textbook or other sources.

_____

_____

_____

_____

_____

5. How do the kidneys show evidence of a Creator?

_____

_____

_____

_____

_____

TABLE 1 Indicator Results

	Albumin (biuret reagent)	Glucose (Benedict's solution)	Starch (iodine solution)	NaCl (multimeter)
Water				
Mixed Solutes Solution				
Beaker A Water				

24A LAB

Sensational!

Exploring the Sensory Organs

We perceive the world around us through our senses. Seeing a flash of green at a traffic light indicates that it is safe to drive through the intersection. The smell of pot roast in the oven tells us that dinner will soon be ready. The sting of a cold winter wind on the cheek suggests that we may need to put on a heavier coat. Although we label some senses major and others minor, the truth is that the absence of any of them puts a person at a disadvantage. In this lab activity you will test some of both the major and minor senses.

? How do my senses help me gather information about my environment?

Equipment

beakers, 250 mL (2)
metal or glass rods (2)
washable ink pen
metric ruler
blindfold

metric tape measure
tuning forks of assorted frequencies (including 128 Hz and 256 Hz)
hot water

ice water
laboratory apron
goggles

PROCEDURE

Sensing Hot and Cold

Our skin receives numerous sensations. The sense of touch involves a particular set of nerves. Pain, temperature, and pressure (and possibly other sensations) are perceived by other nerve endings in the skin. In this portion of the activity you will determine whether hot and cold are different sensations of the same nerve or of two different nerves.

A Obtain a beaker of hot water and place a blunt metal or glass rod in it. (Make sure that the water is not hot enough to burn the skin.) Also obtain a beaker of ice water and place a similar rod in it.

B With a washable ink pen place six small dots on the back of your partner's hand separated by at least 1 cm. Diagram the hand on a piece of paper, draw in the dots, and assign each dot a letter.

C Blindfold your partner.

D Select either a hot or cold rod. After quickly drying the rod, randomly touch the rod to one of the dots on your partner's hand for three seconds. Your partner must then state whether the rod felt hot or cold. Mark a correct response in the appropriate column of Table 2. You do not need to record an incorrect response.

QUESTIONS

- How does the skin sense hot and cold?

- What aspects of vision can I test for?

- How do my ears determine the direction from which a sound is coming?

E Return the rod to its proper beaker between tests to keep it at the correct temperature.

F Repeat Steps D–E until every dot has been tested twice for hot and twice for cold.

1. Which did your partner more often correctly sense, hot or cold?

_____

2. Were each of the locations equally sensitive to hot and cold, or were some more prone to be correct about one temperature or the other?

_____

_____

_____

3. If different areas were sensitive to different temperatures, what might be the reason?

_____

_____

_____

4. If all areas were equally sensitive to different temperatures, what might be the reason?

_____

_____

_____

G With a washable ink pen mark a 2.5 cm square on the back of your partner's hand and then draw a 6 × 6 grid in the square like the ones shown below.

H Use the cold rod to test all the boxes within the grid. Mark in the "Cold" box below all the corresponding areas that were sensitive to the cold rod.

I Repeat Step H for the hot rod. Mark in the "Hot" box below all the corresponding areas that were sensitive to the hot rod.

Cold   Hot

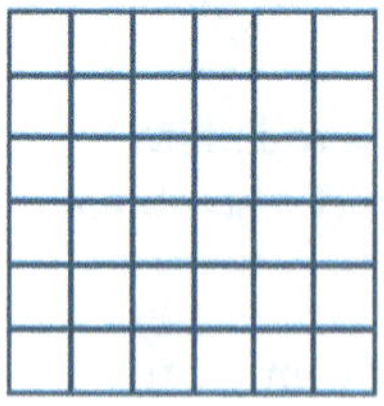

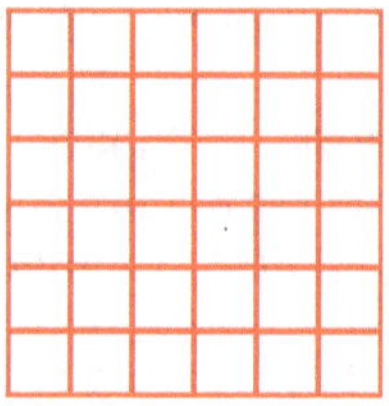

5. According to your experiment, are heat and cold receptors in identical locations on the hand? What can you conclude about temperature receptors?

_____

_____

_____

Vision

Vision is one of the senses on which humans rely most. It is also more prone to minor disorders than are other senses. A slight shortcoming of a person's vision or hearing may go unnoticed for years. For example, astigmatism in one eye may be compensated for by the other eye so that a person may not be aware of having blurred vision. Some people who have red-green colorblindness (one of the most common forms of colorblindness) do not realize they have it until they are adults. In the following activity you will perform some simple tests on your (or your lab partner's) vision. First, you will check yourself for astigmatism.

6. Using a dictionary or internet source, define astigmatism.

_____

_____

_____

J Cover your left eye and hold the image below about 15 cm in front of your right eye. Look directly into the center of the empty circle of the diagram. Be sure that your pupil is looking directly at the central white area. If any of the lines appear blurred, you probably have astigmatism in the corresponding area of your right eye. Repeat the test on your left eye using the same procedure.

7. Do you have astigmatism in your right eye? If so, in which areas (indicated by the numbers at the ends of the lines)?

_____

8. Do you have astigmatism in your left eye? If so, in which areas (indicated by the numbers at the ends of the lines)?

_____

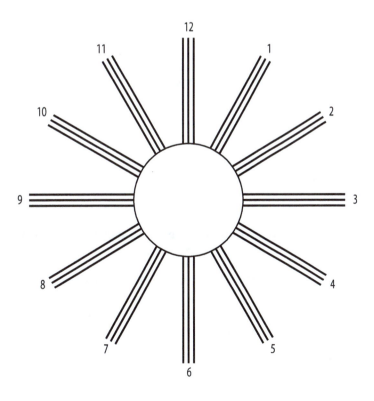

There are no photoreceptors in the eye where the optic nerve fibers leave the eye to form the optic nerve. This area is called the *blind spot*. In people with normal eyes the blind spot of each eye affects a different area of vision; therefore, the total field of vision is unbroken.

K Position the image below so that the dot is to the right of the plus sign. Hold the diagram about 45 cm from your eyes with the plus sign directly in front of your right eye. Cover your left eye.

L Move the diagram slowly toward you as you stare at the plus sign. At a certain point the dot will disappear. The dot is then in your blind spot. Have your partner measure the distance from your eye to the paper.

9. What is the distance from your right eye to the paper at your blind spot?

_____

M Repeat Steps K and L for your left eye.

10. What is the distance from your left eye to the paper at your blind spot?

_____

11. What effect do you think the blind spot would have on a person who has only one eye?

_____

_____

_____

Now let's test your eyes' ability to focus.

N Have your partner look at an object across the room while you hold a pencil about 45 cm in front of his or her nose. Then ask your partner to look at the pencil.

12. How did the pupils change?

_____

13. Did your partner's eyes move in any other way? If so, how?

_____

Another important part of vision is *near-point accommodation*. The near point is the closest distance at which sharp focus is attained.

O Determine the near point of your eyes by covering your left eye and focusing with your right eye on the *D* at the beginning of this sentence. Move this page toward your right eye until the letter no longer appears sharp and clear. Move the page away until the letter is clear again.

P Have your partner measure the distance from the page to your eye to the nearest 0.1 cm.

14. What is your right eye's near-point accommodation distance?

_____

Q Repeat Steps O and P for your left eye.

15. What is your left eye's near-point accommodation distance?

_____

One's near-point accommodation distance increases with age because the lens of the eye loses its elasticity. In the table below notice the average near-point accommodation distance by age.

TABLE 1 Average Near-Point Accommodation by Age

AGE (years)	DISTANCE (cm)
15	8.9
25	11.4
35	17.2
45	52.1
55	83.8

16. Is your near-point accommodation distance normal for your age? Explain.

_____

_____

_____

_____

17. When a person can no longer focus on the print of a book at a "comfortable" distance (about 50 cm), reading glasses are necessary. Considering your present near-point accommodation distance, at what age should you expect to need reading glasses?

_____

Hearing

You depend greatly on your sense of hearing. The ear plays a vital role in sound perception. Your ears tell you more than just the kind and volume of sound, as this lab activity will demonstrate.

The tuning fork is an instrument used to produce a specific pitch. This instrument can be used to test for possible hearing problems. You will use it in the next few experiments. A tuning fork consists of a base and a forked pair of prongs. The base supports the prongs, allowing them to vibrate and produce sound when they are struck.

When using the tuning fork during a test, be careful to hold it at its base only; touching the fork along the prongs interferes with its vibrating and thus stops the sound. Also, strike the fork against the palm of your hand or your thigh only. Do not strike it on any hard surface since resulting dents may alter its frequency.

R Sit in a quiet room with your eyes closed. Your partner will strike the fork on the hand and hold it 20–25 cm away from the sides, top, and then back of your head.

S As your partner moves the fork to different positions around your head, point to the direction from which the sound is coming. Your partner may need to strike the tuning fork each time it is moved.

18. Were you able to locate sound near the sides of your head? the back of your head? the top of your head?

_____

_____

_____

T Place your index finger over the opening of one of your ears and repeat the experiment.

19. Was there any change in the results? Why or why not?

_____

_____

_____

20. In what ways do our senses equip us to be effective servants for God?

_____

_____

_____

TABLE 2 Correct Identification of Temperature for Various Areas of the Hand

	HOT CORRECTLY IDENTIFIED	COLD CORRECTLY IDENTIFIED
Dot A		
Dot B		
Dot C		
Dot D		
Dot E		
Dot F		

24B LAB

Rat Recap

Dissecting a Rat

Rats flourish around human habitation, often to the detriment of people living in the vicinity. Rats eat just about anything and spoil much that they leave behind. They also carry disease and are believed to be partly responsible for the spread of bubonic plague that killed an estimated twenty to fifty million Europeans in the fourteenth century.

But in the last hundred years, domesticated brown rats have been kept as pets and have earned a reputation for being easily trained and loyal to their owners. Other domesticated brown rats have been put to work laying computer cable or serving as therapy animals. Gambian pouched rats have been trained to detect land mines and tuberculosis. But probably the most common use for domesticated rats is as laboratory animals in medical and behavioral research. In this lab activity you will dissect a rat to see the structures typically found in mammals.

Scientists often study mammals in order to learn more about humans. Many medicines are tested in rats or mice to ensure their safety before they are made available for use by humans. Rats also have a body plan similar to that of humans. Think of this activity as the capstone of your study of human anatomy. Obviously, a rat is different from a human, but its internal structure is similar enough that it will allow you to visualize how human internal organs are arranged.

? What typically mammalian structures can be observed in a rat?

Equipment

preserved rat
dissection pan
dissection kit
pins
reference materials

mounted rat skeleton
laboratory apron
nitrile gloves
goggles

QUESTIONS

- What is the best way to dissect a vertebrate?

- How can a rat be used as a model of human anatomy and physiology?

PROCEDURE

Getting Ready

Read all of Lab 24B before you come to class and review Section 19.2 in your textbook. Also feel free to read and consult other sources as you work on the lab activity. These outside sources can help you find information that will be helpful to know to complete this activity. Your teacher can suggest texts for you to consult, and you may also find some information on the internet. A search on rat dissection and rat anatomy may yield good results.

External Structures

A Identify on your specimen each of the structures listed below. Check the box beside each structure after you locate it.

☐ auricles

☐ vibrissae (whiskers)

☐ external nares (nostrils)

☐ upper eyelid

☐ lower eyelid

☐ incisors

☐ forelimbs

☐ hind limbs

☐ head

☐ trunk

☐ anus

☐ tail

☐ (male:) penis and scrotal sacs (containing testes)

☐ (female:) vaginal opening and teats

☐ fur

1. Do the auricles contain bone?

_____

2. What is the function of the vibrissae?

_____

_____

3. Why do rats have such prominent incisors?

_____

_____

4. Of the upper and lower incisors, which ones are longer?

_____

5. How is a rat's upper lip different from that of a human?

_____

6. How is the surface of the tail different from the rest of the body?

_____

7. How many toes does a rat have on each of its forelimbs? on each of its hind limbs?

_____

8. What is the sex of your rat?

_____

Removing the Skin

Now you will remove the rat's skin.

B Place your rat in the dissection pan with its ventral side up.

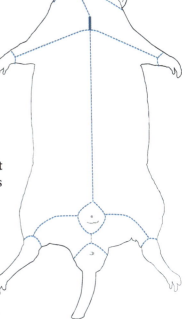

C Pinch the skin in the thoracic region (over the sternum) enough to puncture it with the tip of your scissors. Carefully cut by continuing to raise the skin from the muscles and cut along the indicated lines with your scissors, penetrating only into the subcutaneous layer (see right). Do not cut into the muscles because you will need to observe them intact. Note that you will cut around each limb just proximal to each foot. You will also cut around the neck, the base of the tail, and the groin area, leaving the fur intact.

D After you have made all the indicated cuts, use your forceps to pull the skin away from the muscles beneath. It should peel back as flaps if you carefully use your scissors and perhaps the probe to separate the skin from the muscles.

E Turn your rat over so that the dorsal side is up. Now complete the skin removal by pulling the skin away from the back in one piece. If all cuts were made properly, this should be possible.

F Examine the skin that you have removed.

9. How many layers can you see?

_____

10. How many types of hair can you discern?

_____

Muscles and Bones

Now that you've skinned your rat, look for the muscle groups that you learned for humans in Chapter 21. Muscle functions in the rat are often different from those in the human because of the difference in body plans. The origins and insertions, though, are generally the same and are summarized in Table 1 (p. 262). You should be able to see all the muscle groups indicated without making any further cuts. You will probably be able to distinguish the superficial muscles shown on the next two pages by carefully separating them with your probe. Simply insert the probe under the loose edge of the muscle and slide along its margin. You can usually tell where one muscle separates from another by the direction of the fibers.

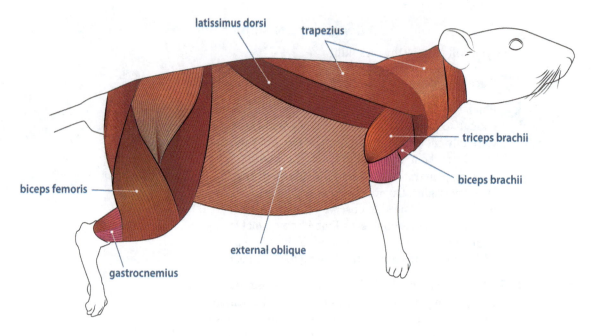

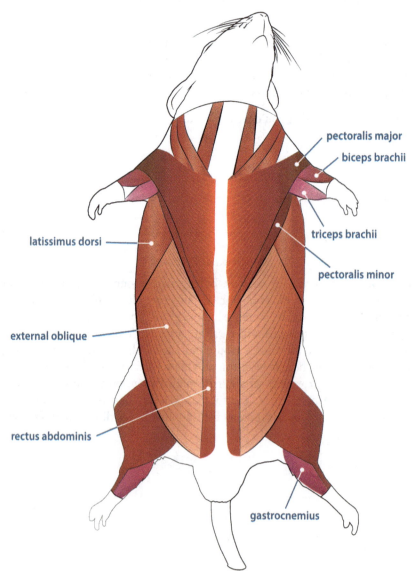

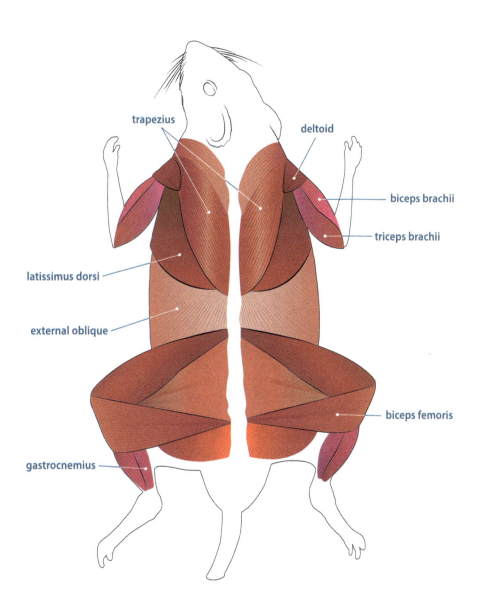

trapezius

deltoid

biceps brachii

triceps brachii

latissimus dorsi

external oblique

biceps femoris

gastrocnemius

G As you find each muscle or muscle group, check the box beside its name in Table 1. After mastering the identification of the muscles in the chart and the figures, choose one hind leg and carefully remove the biceps femoris, gastrocnemius, and any other necessary underlying muscles to reveal the tibia, fibula, and femur, as well as the patella.

TABLE 1 Muscles of the Rat

NAME	ORIGIN	INSERTION	FUNCTION
☐ trapezius	vertebrae	scapula	pulls scapula inward, up, and backward
☐ latissimus dorsi	vertebrae	humerus	moves humerus toward the back
☐ external oblique	muscles in region of last nine to ten ribs	pelvis	compresses abdomen
☐ gastrocnemius	lower femur	Achilles tendon	extends the foot
☐ biceps femoris	lower vertebrae	femur and tibia	flexes the lower rear leg
☐ biceps brachii	scapula	radius	flexes lower front leg
☐ triceps brachii	scapula and humerus	ulna	extends lower front leg
☐ deltoid (spinodeltoid)	spine near the scapula	humerus	pulls scapula back and outward
☐ pectoralis major	sternum	humerus	draws front leg forward, toward chest
☐ pectoralis minor	sternum	humerus and scapula	draws front leg forward, toward chest
☐ rectus abdominis	pelvis	sternum	compresses abdomen

Many of the bones in the human body can be found in a rat, though they may look somewhat different because of the different body plans that God has designed.

H Examine your skinned specimen to identify the bones shown below. As you find each one, check the box by its name.

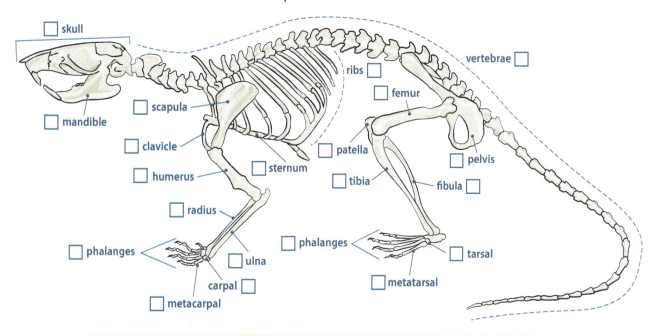

☐ skull

☐ mandible

☐ scapula

☐ clavicle

☐ humerus

☐ radius

☐ phalanges

☐ sternum

☐ ulna

☐ carpal

☐ metacarpal

☐ ribs

☐ patella

☐ tibia

☐ phalanges

☐ metatarsal

☐ vertebrae

☐ femur

☐ pelvis

☐ fibula

☐ tarsal

Opening the Body Cavity

There are many ways to open the body cavity of a rat. The method described below is recommended. It usually results in fewer damaged organs.

I Place your rat in the dissection pan with its ventral side up.

J Use your scissors to cut through just the muscle layers along the lines indicated on the right. It is best to make the longitudinal cut first and then the shorter transverse cuts. Be careful not to cut the internal organs as you make the incisions.

K After all the cuts are made, fold the flaps back and pin them to the pan so that the organs are more easily seen. Do not remove limbs or any other structures unless your teacher instructs you to do so.

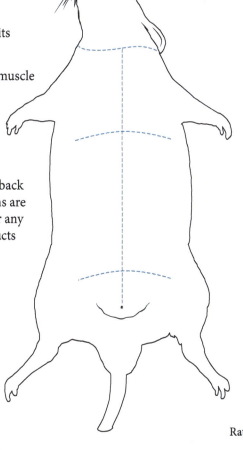

THE CIRCULATORY AND LYMPHATIC SYSTEMS

L Identify the labeled structures below. As you find each one, check the box beside its name. Mesenteries (thin, clear connective tissues with blood vessels running through them) hold the organs in place. Using a probe to loosen these mesenteries will make the internal organs more visible.

Note: If your rat is older, the thymus may not be distinguishable.

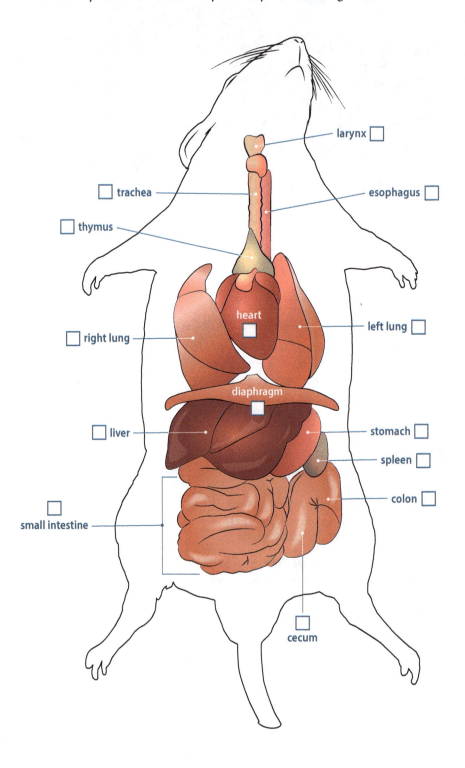

M While the rat's heart is in its natural position, identify the parts of the heart listed on the right. Check the box by each structure as you find it.

11. Which structures of the heart contain deoxygenated blood?

_____

N Remove the heart.

THE RESPIRATORY SYSTEM

O Remove the respiratory system. Follow the trachea as near to the mouth cavity as possible. Include the larynx when you make the cut to remove this system.

12. How do the rat's diaphragm and rib cage function in breathing?

_____

_____

_____

_____

_____

_____

left atrium ☐

right atrium ☐

right ventricle ☐

left ventricle ☐

THE DIGESTIVE SYSTEM

P Remove the structures of the digestive system by carefully cutting the esophagus as close to the skull as possible and cutting across the rectum, posterior to the end of the colon.

Q Carefully separate any mesenteries connecting digestive organs to parts of the urogenital system and the vertebrae. The entire digestive system—with the liver, spleen, and pancreas intact—should now come out of the body cavity.

R Slowly tease the mesenteries with your probe so that the alimentary canal can be straightened. All accessory organs should remain attached to the alimentary canal.

13. (a) How long is the alimentary canal? (b) How long is it in relation to the length of the rat?

_____

_____

14. In regard to the rat and other vertebrates, what is the relationship between an animal's diet and the length of the intestine?

_____

_____

15. How many lobes does the liver have?

_____

THE URINARY, ENDOCRINE, AND REPRODUCTIVE SYSTEMS

These structures in a rat are basically the same as in a human.

S Carefully locate the male or female urinary structures shown below and check the box beside the name of each structure as you find it. Do not remove the urinary structures from the rat body cavity.

The ureters are sometimes hard to find because of their small size, but if you wiggle the kidneys it may be easier to see them. Note that the adrenal glands, atop the kidneys, are considered part of the endocrine system.

T Carefully locate and expose the male or female reproductive structures shown below. If you are dissecting a male rat, you will need to cut the scrotal sac to access the testes. Leave them in the rat body cavity. As you find each structure, check the box beside its name.

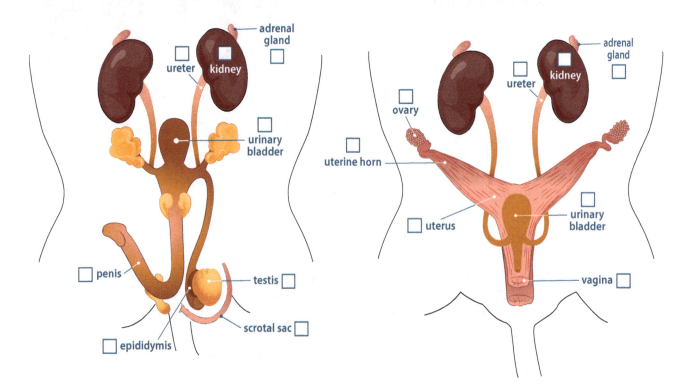

Male Rat Urogenital System

- adrenal gland
- ureter
- kidney
- urinary bladder
- penis
- testis
- scrotal sac
- epididymis

Female Rat Urogenital System

- adrenal gland
- ureter
- kidney
- ovary
- uterine horn
- uterus
- urinary bladder
- vagina

25A LAB

Unusual Development

Modeling the Amazing Growth of Robert Wadlow

You've learned that as you get older, you get taller. But you've also learned that at some point you stop getting taller even though you're getting older. It's a good thing too, otherwise you might be 16 feet tall when you retire!

Have you ever wondered how tall the tallest person in the world was? Did he grow like a normal person? If not, how was his growth different?

The famous Guinness Book of World Records has verified Robert Pershing Wadlow to be the tallest known human in history. Robert was born in 1918 at a normal baby weight of about 4 kg. His record-breaking growth soon began. When Robert was twelve years old, doctors concluded that he had an enlarged, hyperactive pituitary gland that produced abnormally high levels of growth hormone. Robert passed the height of an average adult male by the age of seven. His four younger siblings all matured to a normal size, and there is no record of unusual growth for any other family members. Average males stop growing at about age eighteen, but Robert continued growing throughout his life.

In this lab activity you will graph and compare data from Robert Wadlow's amazing growth with that of a normal male to see how different Robert's growth was.

? **How does the growth of the tallest man to ever live compare with normal male development?**

Equipment

colored pencils, 4 different colors
calculator

QUESTIONS

- What is the relationship between Robert Wadlow's growth and normal growth?
- How can you interpret trends in data?
- Can data be extrapolated from existing information?

PROCEDURE

Graphing the Data

Let's see what you can learn from Robert's incredible growth. You will be using the data in Table 1 (see next page) to create two "dual-line" graphs.

A Create a graph of male height versus age for both Wadlow and a normal male in Graphing Area A. Be sure to label the resulting lines.

B Create a graph of male mass versus age for both Wadlow and a normal male in Graphing Area B. Be sure to label the resulting lines.

TABLE 1 Growth Data for Robert Wadlow and Normal Males

| Age (y) | ROBERT WADLOW'S GROWTH | | NORMAL MALE GROWTH | |
	Height (cm)	Mass (kg)	Height (cm)	Mass (kg)
Birth	51.0	3.8	49.8	3.3
1	107.0	20.0	75.7	9.6
2	137.0	34.0	86.8	12.5
3	150.0	40.0	95.2	14.0
4	160.0	48.0	102.3	16.3
5	169.0	64.0	109.2	18.4
6	170.0	66.0	115.5	20.6
7	178.0	72.0	121.9	22.9
8	183.0	77.0	128.0	25.6
9	188.0	82.0	133.3	28.6
10	196.0	95.0	138.4	32.0
11	211.0	109.0	143.5	35.6
12	218.0	130.0	149.1	39.9
13	224.0	137.0	156.2	45.3
14	226.0	150.0	163.8	50.8
15	239.0	161.0	170.1	56.0
16	248.0	170.0	173.4	60.8
17	251.0	173.0	175.2	64.4
18	254.0	177.0	175.7	66.9
19	259.0	220.0	176.5	68.9
20	262.0	221.0	177.0	70.3
21	264.0	223.0		
22.4*	272.0	199.0		

*Age of Wadlow at his death

CONCLUSIONS

Answer the following questions on the basis of your graphs and the data in Table 1.

1. In what year did Wadlow's height increase the most? When does the average male grow the most?

   _____

2. What was the average amount that Robert grew in height per year? What is the average amount that the average male grows in height in one year?

   _____

3. How does Robert's increase in height compare to other males?

   _____

   _____

4. In what year did Wadlow's mass increase the most? When does the average male increase in mass the most?

   _____

5. What was the average amount that Robert increased in mass per year? What is the average amount that the typical male increases in mass per year?

   _____

6. How does Robert's growth in mass compare to other males?

   _____

   _____

7. Scientists often extrapolate data to make predictions. To do this, they extend a graph along the same slope, above or below the measured data. If Robert had lived until the age of thirty and had continued growing at the same rate that you previously calculated, what height might he have attained? What mass might he have grown to?

   _____

The formula for calculating BMI—body mass index, a calculation used to assess the relationship between a person's height and mass—is

$$weight\ in\ kg \div (height\ in\ cm)^2 \times 10\,000.$$

8. According to the US Centers for Disease Control, a healthy BMI is 18.5–24.9 kg/cm². Calculate Robert's BMI at the ages of five, ten, fifteen, and twenty. Would Robert have been considered healthy?

   _____

   _____

   _____

9. After consulting the internet or other resources, list some other health problems that might be faced by a true "giant" other than the ones mentioned in this lab activity.

_____

_____

_____

10. What kinds of additional modifications might a person of this size need for daily life?

_____

_____

_____

GOING FURTHER

Most pediatricians do a rough estimate of how tall a boy will grow using the formula shown below.

(mother's height + 5 in. + father's height)/2

Robert's father was 5 ft 11 in. tall. There is not an easily obtainable recorded height for his mother, but by comparing photos, we can surmise that she was about 5 ft 5 in. tall.

11. Using these measurements of Robert's parents, how tall would a pediatrician predict one of their sons to be?

_____

12. What is the significance of the pituitary gland?

_____

_____

13. What might cause the pituitary to become enlarged?

_____

_____

14. If Robert Wadlow had been born in 2020, how might doctors have treated him?

_____

_____

15. Considering the health issues associated with acromegaly, how does the function of the pituitary gland demonstrate the wisdom of God's design for the human body?

_____

_____

_____

GRAPHING AREA A

GRAPHING AREA B

25B LAB

Fast Food Fact-Finding

Exploring the Perception of Fast Food versus Reality

Fast food is big business! In 2015 estimated US fast food sales generated nearly five times the amount of revenue earned by movies and professional sports combined. In fact, during that same year Americans spent an amount of money on fast food greater than the gross domestic products of three-quarters of the world's countries. That's a lot of money—and a lot of food! Yes, American fast food is popular, both here and around the world, but you would probably have to be a hermit not to have heard some of the troubling accusations being made about fast food. Various studies, news outlets, and talk-show doctors have linked fast food to obesity, heart disease, diabetes, and a host of other maladies. But is fast food really that bad for you? In this activity you will examine fast food, and you'll make use of a surprising source of information—the fast food franchises themselves.

Is eating fast food a healthy lifestyle option?

?

Equipment

PROCEDURE

Road Trip!

Imagine this scenario: you just got your driver's license, and you've been given use of the family car for the whole day! So you head out early, deciding that you'll stop somewhere along the way for meals. With no one along on the trip to tell you something's bad for you or costs too much, you get to eat whatever you want at whatever restaurants you want. Ah, but here's the catch:

» You'll need to eat three meals (breakfast, lunch, and dinner) plus dessert.

» For each meal you'll need to record the name of the restaurant where you eat. You must include an itemized list of everything you eat and drink at that restaurant, using the menu names as given by the restaurant. If you order a combo, you'll need to list separately everything included in the combo.

» The only restriction on where you can eat is that it must be a restaurant that publishes nutrition information that you can access either at the restaurant, in a brochure, or online.

Don't overthink what you'll order. Remember, it's your road trip, so order whatever you want.

QUESTIONS

- What are my fast food preferences?
- Do my fast food choices increase my health risks?
- Do I need to change my fast food eating habits?

A Record what you ate at each meal in the Menu Item column of Tables 1–4.

1. Think about the nutrients that you've learned about this year. Describe whether you feel that the meals you've chosen for the day will satisfy your body's nutritional needs.

   _____

   _____

   _____

   _____

Breaking Down the Nutrition Information

Most restaurants now post nutritional information for their menus online. To find the nutritional information for each restaurant that you visited on your road trip, do an internet search for the name of the restaurant plus the keyword "nutrition."

B Consult each restaurant's online nutrition information, then fill in the number of calories and the amount of fat, carbohydrates, protein, and sodium for each item that you listed in Tables 1–4.

C Add up the total amount of each nutrient and record the results in the last row of each table.

D Now add up the totals from each meal to find the total amount of each nutrient consumed during the day and record the results in Table 5.

Measuring Up

Now let's find out whether your day of dining out on the road provided you with adequate nutrition. First, you'll need some idea of what the basic nutritional needs of a teenager are. Not all sources agree on the nutritional needs of teenagers, but for this lab activity we'll use the guidelines published by the US Department of Agriculture (USDA) for 2015.

E Consult Table 6, the USDA Estimated Calorie Needs per Day by Age, Sex, and Physical Activity Level, to determine your daily calorie need.

2. What is your daily calorie need?

   _____

3. Express your caloric intake for the day from Table 5 as a percentage of your estimated calorie need from Table 6.

   _____

4. How does your caloric intake for the day compare to the calorie need suggested by the USDA?

   _____

   _____

Table 7 shows the USDA recommended daily intake amounts for teens for several nutrients.

5. How does your intake for the day for each of these nutrients compare to the recommended amounts in Table 7?

_____

_____

_____

_____

Assessing Risk

6. Considering your road trip data and the USDA recommended calorie needs for teens, do you think eating fast food increases your risk for weight gain? Explain.

_____

_____

_____

_____

Type 2 diabetes is a disease that results in the body being unable to properly use and store sugars and fats. There is no cure for type 2 diabetes. Once a person has diabetes, the disease must be managed for the rest of his or her life. One of the risk factors for developing type 2 diabetes is being overweight.

7. Considering your road trip data, do you think eating fast food increases a person's risk of developing type 2 diabetes? Explain.

_____

_____

_____

_____

As you learned in Lab 22B, hypertension (high blood pressure) is a risk factor for stroke. Your chances of developing hypertension are increased by being overweight and by eating a high-sodium diet.

8. Considering your road trip data, do you think eating fast food increases a person's risk of developing hypertension?

_____

_____

_____

_____

The risk factors for heart disease include being overweight and having hypertension, diabetes, or both.

9. Considering your road trip data, do you think eating fast food increases a person's risk of developing heart disease?

_____

_____

_____

_____

Think about It

10. Suggest some alternatives to eating fast food during your road trip.

_____

_____

_____

_____

11. Are there ways to eat healthier even at a fast food restaurant? Explain.

_____

_____

_____

_____

12. Fast food advertising often attempts to portray fast food as fun, tasty, convenient, and inexpensive—just think about the commercials you have seen on TV showing fit young people or families smiling and enjoying a hamburger or fried chicken strips. Suggest a more balanced perception of fast food.

_____

_____

_____

_____

_____

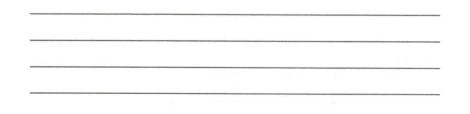

One slice: 435 Cal, 25 g of fat, and 883 mg of sodium

13. Why should Christians give careful consideration to their fast food dining habits?

_____

_____

_____

_____

_____

_____

_____

_____

_____

_____

GOING FURTHER

14. Obesity, diabetes, hypertension, and heart disease are only a few of the problems that have been linked to fast food. Do an internet search on the topic of fast food health risks, then briefly discuss some of the additional risks associated with fast food.

_____

_____

_____

_____

_____

_____

_____

TABLE 1 Breakfast

NAME OF RESTAURANT: _____

Menu Item	Calories (kcal)	Fat (g)	Carbohydrates (g)	Protein (g)	Sodium (mg)
Totals					

TABLE 2 Lunch

NAME OF RESTAURANT: _____

Menu Item	Calories (kcal)	Fat (g)	Carbohydrates (g)	Protein (g)	Sodium (mg)
Totals					

TABLE 3 Dinner

	NAME OF RESTAURANT: _____				
Menu Item	Calories (kcal)	Fat (g)	Carbohydrates (g)	Protein (g)	Sodium (mg)
Totals					

TABLE 4 Dessert

	NAME OF RESTAURANT: _____				
Menu Item	Calories (kcal)	Fat (g)	Carbohydrates (g)	Protein (g)	Sodium (mg)
Totals					

TABLE 5 Daily Total

Calories (kcal)	Fat (g)	Carbohydrates (g)	Protein (g)	Sodium (mg)

TABLE 6
USDA Estimated Calorie Needs per Day by Age, Sex, and Physical Activity Level

	MALES				FEMALES		
Age	Sedentary	Moderately Active	Active	Age	Sedentary	Moderately Active	Active
14	2000	2400	2800	14	1800	2000	2400
15	2200	2600	3000	15	1800	2000	2400
16	2400	2800	3200	16	1800	2000	2400

Sedentary means a lifestyle that includes only the physical activity of independent living. *Moderately Active* means a lifestyle that includes physical activity equivalent to walking about 1.5 to 3 miles per day at 3 to 4 miles per hour, in addition to the activities of independent living. *Active* means a lifestyle that includes physical activity equivalent to walking more than 3 miles per day at 3 to 4 miles per hour, in addition to the activities of independent living.

TABLE 7 USDA Nutritional Goals

MACRONUTRIENT OR MINERAL	MALE 14–18	FEMALE 14–18
Fat (g)*	70–100	55–75
Carbohydrates (g)	130	130
Protein (g)	52	46
Sodium (mg)	2300	2300

* Fat amounts are calculated according to recommended caloric intake of fats for moderately active teens.

APPENDIX A

Laboratory and First Aid Rules

Laboratory Rules

1. Never perform an unauthorized experiment or change any assigned experiment without your teacher's permission.

2. Avoid playful, distracting, or boisterous behavior.

3. Never work alone. Students must not conduct lab activities without supervision.

4. Work at your own lab station.

5. Always work in a well-ventilated area. Use the fume hood when working with toxic vapors. Never put your head in the fume hood.

6. Always wear safety goggles when working with chemicals, glassware, projectiles, and other materials or objects that are potentially hazardous to the eyes.

7. Wear protective clothing and gloves when working with corrosive or staining chemicals.

8. While working in the laboratory, tie back long hair and avoid wearing loose clothing such as scarves or ties.

9. Never taste any chemical, eat or drink out of laboratory glassware, or eat or drink in the laboratory.

10. Always use appropriate instruments for cutting and handle them carefully. Always cut away from yourself.

11. When handling live organisms, follow instructions and do not cause them undue harm or discomfort.

12. Thoroughly wash your hands with soap after handling any live organisms, cultures containing organisms, or chemicals.

13. To smell a substance, gently fan its vapor toward you.

14. Never leave a flame or heater unattended. Keep combustible materials away from heat sources.

15. When diluting acid solutions, always add the acid to water slowly. ***Never add water to an acid!***

16. When heating a test tube, point the open end away from you. ***Never heat a closed or stoppered container!***

17. Dispose of waste as instructed by your teacher.

18. Do not return unused chemicals to a container. Dispose of them properly.

19. Notify the teacher of any injuries, spills, or breakages.

20. Know the locations of the fire extinguisher, safety shower, eyewash station, fire blanket, first-aid kit, and Safety Data Sheets (SDSs).

First Aid Rules

BURNS

Flush the area with cold water for several minutes. Do not apply ice.

CHEMICAL SPILLS

Notify your teacher of all chemical spills.

ON A LABORATORY DESK

1. If the material is not particularly volatile, toxic, or flammable, your teacher may have you clean the spill. For liquids, use an absorbent material that will soak up the chemical. For solids, use the designated dustpan and brush. Dispose of chemicals and cleaning materials properly. Then clean the area with soap and water.

2. If the material is volatile, toxic, or flammable, and not a large spill, ask your teacher for help. If it is a large spill, you may need to evacuate the laboratory.

3. If a highly reactive material, such as hydrochloric acid, is spilled, your teacher will clean it up.

ON A PERSON

1. If the spill covers a large area, begin rinsing in the chemical shower, then remove all contaminated clothing and remain under the safety shower. Flood the affected body area for fifteen minutes. Obtain medical help immediately.

2. If the spill covers a small area, immediately flush the affected area with cold water for several minutes. Then wash the area with soap and water. Get medical attention.

3. If the chemical splashes in the eyes, immediately wash them in the nearest eyewash fountain for at least fifteen minutes. Get medical attention.

4. If the spill is an acid, rinse the affected area with sodium bicarbonate or sodium carbonate solution; if it is a base, use citric or ascorbic acid solution.

FIRE

1. For any fire other than a contained fire, do *not* attempt to put it out on your own. ***Understand that some fires can't be put out with water.***

2. Smother a small fire in a container by covering it.

3. If a person's clothes are on fire, remember to stop, drop, and roll—roll the person on the floor and use a fire blanket to extinguish the flames. The safety shower may also be used. ***Do not use a fire extinguisher!***

SWALLOWING CHEMICALS

Determine the specific substance ingested and follow the instructions on the SDS. Contact the Poison Control Center in your area immediately.

CUTS

If the wound is superficial, clean it with a disinfectant, apply triple antibiotic ointment, and cover with a bandage. If it is a deep cut, apply pressure and seek immediate medical attention.

BITES AND STINGS

For minor bites (that do not break the skin) and stings, wash the affected area with soap and water and cover with antibiotic ointment and a clean bandage. Communicate with your teacher about any allergies that you may have and whether you have an up-to-date tetanus shot.

If the bite is from an animal, wash the affected area with soap and water and cover with antibiotic ointment and a clean bandage. See a doctor immediately.

RASHES FROM PLANTS

Rashes caused by exposure to poison ivy, oak, or sumac will not appear immediately after contact; it usually takes about twenty-four hours for symptoms to show. If you suspect that you have contacted these plants, wash the area with soap and water within ten minutes of contact. Be sure to clean any clothing that has contacted plants since it may also contain the toxins.

If a rash is visible, treat with calamine lotion, hydrocortisone cream, or oral antihistamines. If conditions worsen, see a doctor.

Laboratory Equipment

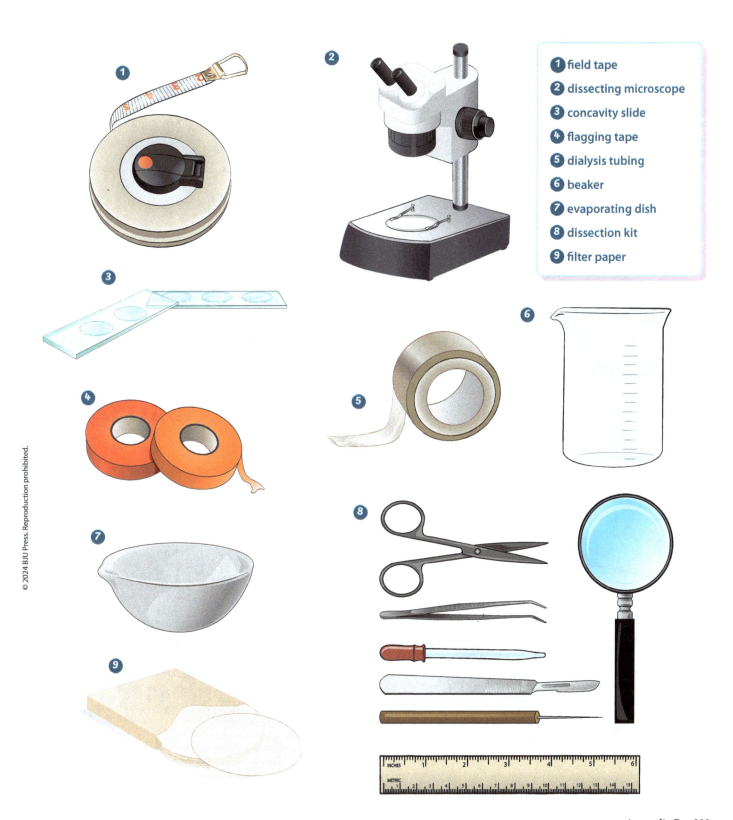

1. field tape
2. dissecting microscope
3. concavity slide
4. flagging tape
5. dialysis tubing
6. beaker
7. evaporating dish
8. dissection kit
9. filter paper

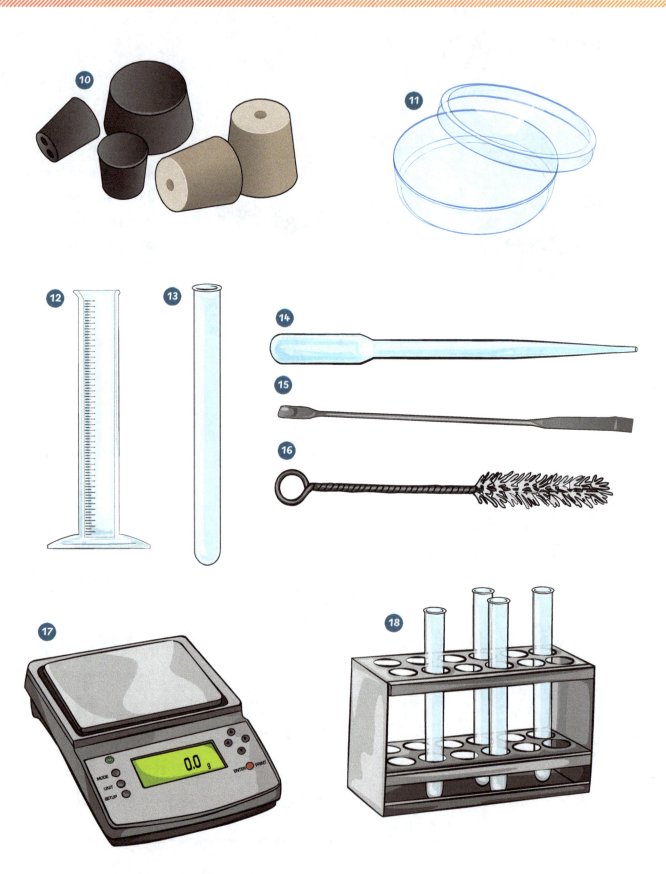

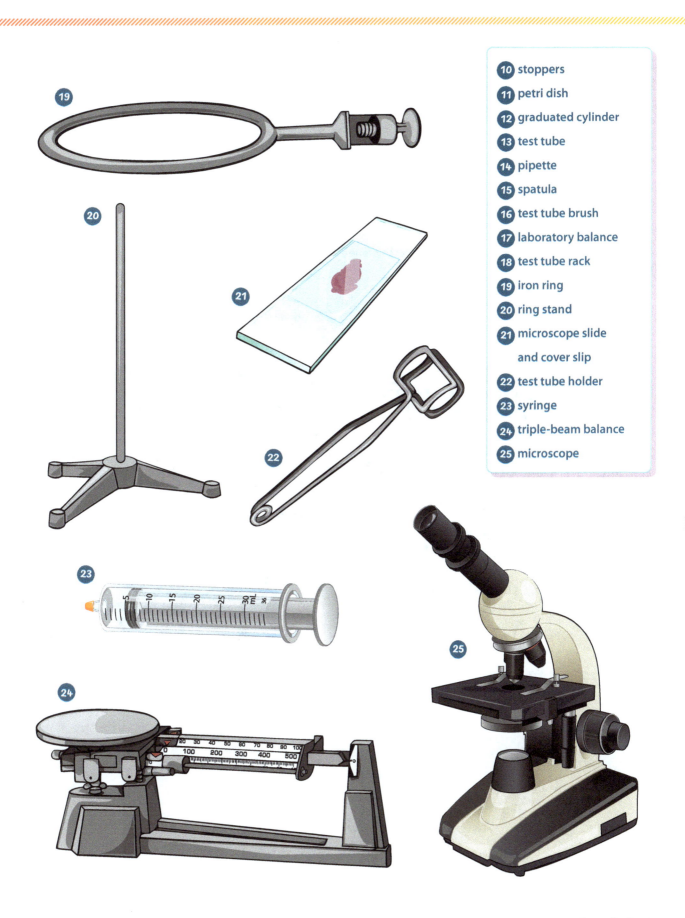

10 stoppers
11 petri dish
12 graduated cylinder
13 test tube
14 pipette
15 spatula
16 test tube brush
17 laboratory balance
18 test tube rack
19 iron ring
20 ring stand
21 microscope slide
   and cover slip
22 test tube holder
23 syringe
24 triple-beam balance
25 microscope

Anatomical Terms

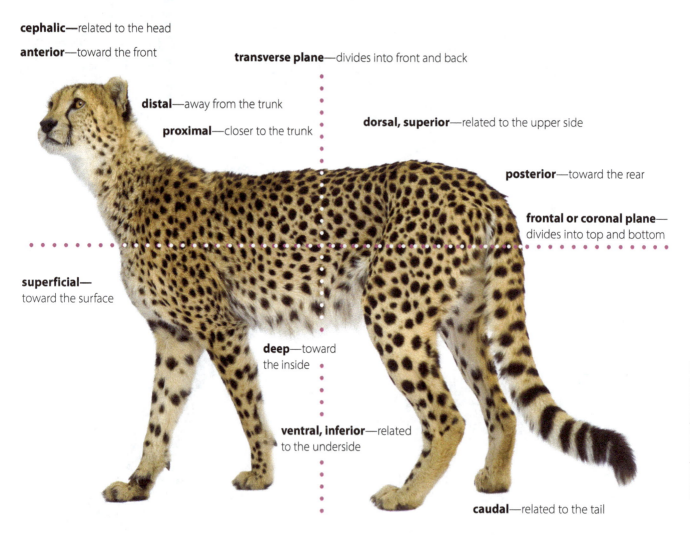

cephalic—related to the head

anterior—toward the front

transverse plane—divides into front and back

distal—away from the trunk

proximal—closer to the trunk

dorsal, superior—related to the upper side

posterior—toward the rear

frontal or coronal plane—divides into top and bottom

superficial—toward the surface

deep—toward the inside

ventral, inferior—related to the underside

caudal—related to the tail

midline (also median or sagittal plane)—divides into left and right

dorsal, superior—related to the top

lateral—related to the side

right side of organism

left side of organism

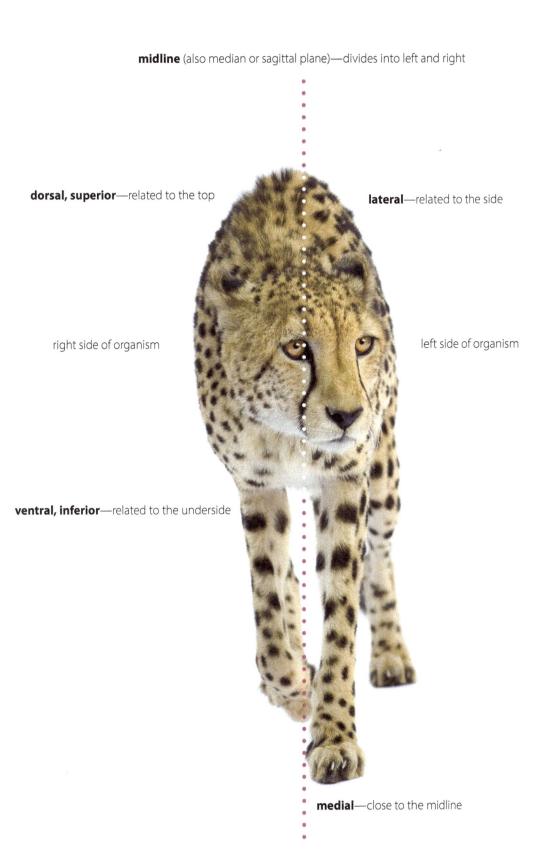

ventral, inferior—related to the underside

medial—close to the midline

APPENDIX D

Laboratory Techniques
Making Biological Drawings

During some lab activities you will be asked to draw what you observe. Drawing scientific specimens is one of the best ways to learn some of the complex biological structures and processes that you will be observing. Drawing is also a great field skill to have. As you draw, concentrate on the shape and textures of the different structures you observe. By the time you have finished, you should better understand the structures that you have drawn.

Drawing areas have been provided for you. But your teacher may also allow you to take a photo, print it out, and include it in the drawing area.

A Center the name of the specimen above the drawing. If the drawing is of a portion of an organism, indicate the portion in the title or under the title. For example, "Fruit Fly, leg" indicates that the drawing is of the leg of a fruit fly. The title "Fruit Fly" alone indicates a drawing of the entire fly.

B If the specimen or structure has been prepared as a wet mount or a longitudinal mount before you observe it, indicate this under the name of the specimen.

C Make the drawing large and center it in the space available.

D You can use just a pencil for most drawings so that you can erase anything you want to change. If it makes your drawing clearer, use colored pencils, though a colorful drawing should not be a goal in itself.

E Add labels after the drawing is complete; label what you can identify and only what you observe.

F If you used a microscope, indicate the power in the lower right-hand corner. If you used some other type of magnification (such as a hand lens), write "magnified" or "enlarged" in that corner. Don't draw the microscope field or the container that the specimen is in.

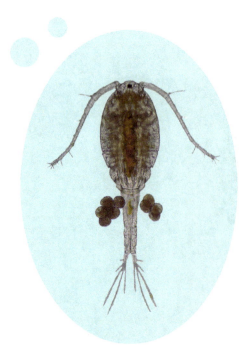

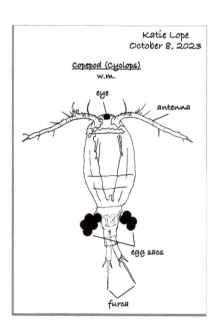

Preparing a Wet Mount

When you use a microscope to look at specimens, sometimes you will use slides that have been prepared ahead of time. But for some lab activities you will need to make your own temporary slide. Placing a specimen in a drop of water to observe it is called *making a wet mount*. To make a wet mount, you need a microscope slide and a cover slip as well as the specimen that you want to observe.

A Handle glass slides and cover slips by their edges so that you do not leave fingerprints on them.

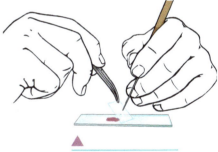

◀ Handling glass slides

B Be careful not to bend plastic cover slips. If you are using a glass cover slip, handle it gently; splinters from shattered glass cover slips easily enter the fingers and may require surgical removal. Inspect plastic cover slips for excessive scratches. If too many scratches appear, discard the cover slip.

C Use only water when washing the slides and cover slips. Soap film may kill or damage living specimens. Shake off excess water and dry the slide with a tissue.

D Place a drop of water on the clean slide.

E Using forceps, place the specimen in the drop of water. In some cases your teacher may provide a special concavity slide that has a shallow well to hold larger specimens.

F Place the cover slip on top of the specimen so that one edge is touching the slide and the cover slip is held at a 45° angle above the drop of water. Slowly lower the cover slip down on top of the water and specimen. If bubbles appear in the area that you are going to view, tap the cover slip with the tip of a probe to remove the bubbles.

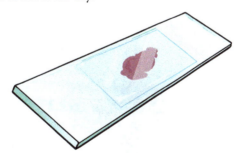

Creating a wet mount

G When finished observing, dispose of the specimen and wash the slide and cover slip, placing them on a paper towel to air dry.

APPENDIX D

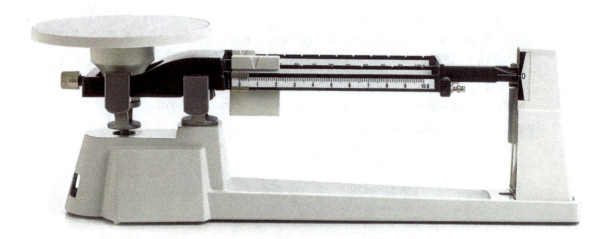

Using Mechanical Balances

The mass of a substance can be determined in the laboratory with the use of a mechanical balance. Several kinds of mechanical balances are common, but all of them operate on the same principles. To use a mechanical balance properly, follow the steps given below.

A Place the balance on a smooth, level surface.

B Keep the balance pan(s) clean and dry. Never put chemicals directly on the metal surface of the pan(s). Always place materials on a sheet of weighing paper, on a weighing boat, or in a container.

C Check the rest point of the empty balance. To do this, remove all weight from the pans and slide all movable masses to their zero positions. If the balance beam swings back and forth, note the central point of the swing. You do not have to wait until the beam stops swinging completely. If the central point on the balance arm does not align with the marked zero point on the post, have your teacher adjust the balance. *Do not adjust the balance yourself!*

D Place the substance on the pan and adjust the sliding masses. Move the largest mass first, and then make final adjustments with the smaller masses. The sum of all the readings is the mass of the object.

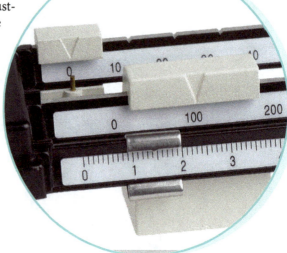

Using an Electronic Balance

Electronic balances are generally faster and easier to use than their mechanical counterparts. To use an electronic balance properly, follow the instructions given below.

A Place the balance on a smooth, level surface.

B Keep the balance pan(s) clean and dry. Never put chemicals directly on the metal surface of the pan(s). Always place materials on a sheet of weighing paper, on a weighing boat, or in a container.

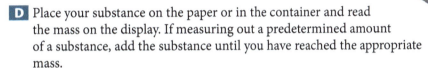

C Turn on the balance. Place the container or weighing paper that will hold the substance whose mass you are trying to find and make sure that there is a reading of 0 by pushing the *Tare* button.

D Place your substance on the paper or in the container and read the mass on the display. If measuring out a predetermined amount of a substance, add the substance until you have reached the appropriate mass.

Using a Thermometer

When using a thermometer in lab activities, make sure that you are using one that has the proper temperature range for the experiment that you will be doing. Support the thermometer in a one-hole rubber stopper when necessary to contain the substance whose temperature the thermometer is measuring. To avoid breaking the thermometer and cutting your hand while inserting it in the stopper, lubricate the thermometer and the stopper hole with soap or glycerol. Then protect your hands with puncture-resistant gloves or several layers of paper towels. Hold the thermometer near the stopper and gently twist it into the hole. If you have to use a great amount of force, ask your teacher to enlarge the hole.

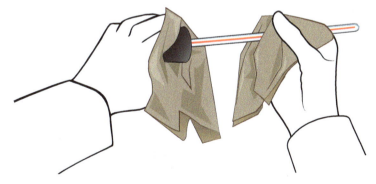

Position the thermometer bulb just above the bottom of the container. If the bulb touches the container, your readings will be inaccurate. Always hold or secure a thermometer in a container; never leave an unsecured thermometer sticking out the top of a container. If a thermometer breaks, alert your teacher and do not touch the inner contents. Some thermometers still in use contain mercury. The spilled mercury may look fascinating, but it is toxic and can be absorbed through the skin.

APPENDIX D

Handling Liquids

Proper technique for handling liquids is essential if you are to remain safe, keep reagents pure, and obtain accurate measurements. For increased safety do not splash or splatter liquids when pouring. Pour them slowly down the insides of test tubes and beakers. If anything is spilled, wipe it up quickly. (See First Aid Rules on pages 281–82.)

To keep the liquid chemicals pure, keep stirring rods out of the stock supply. Do not let the stoppers and lids become contaminated while you are pouring. Instead, hold the stopper between your fingers. If you must put a lid down, keep the inside surface from touching the surface of the table.

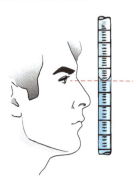

Accurate measurements of liquids can be made in burettes, graduated cylinders, and volumetric flasks. You should measure volumes in these pieces of glassware unless you need only a rough approximation. When reading the level of a liquid, look at the meniscus (curved surface) at the center of the upper surface of the liquid along a horizontal line of sight. For most liquids you will measure at the bottom of the meniscus.

*QUICK NOTE
Measuring Out Powdered Solids

1. Scoop out a little of the sample with a spatula.

2. Gently tap the spatula until the desired amount falls off onto a sheet of weighing paper.

3. Cup the paper to pour the powdered solid into a test tube or other container.

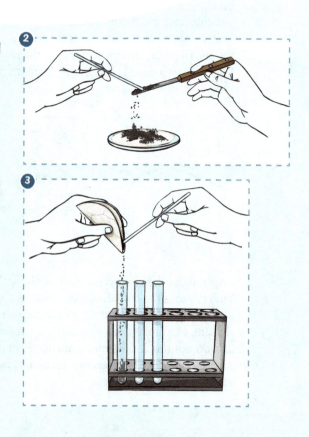

Slowing Protozoan Movement

In some lab activities you will be working with protozoans—microscopic organisms that move. This makes them hard to observe! You can slow down protozoans and make them easier to observe in a wet mount by using a cover slip, cotton fibers, or a thicker medium.

As the culture medium of a wet mount evaporates, the cover slip will press on the organism and slow its movement. Use a paper towel on the edge of the cover slip to speed the process. Your lab partner can draw off small portions of water while you chase the organism. Do not permit the medium to evaporate completely. Replenish it by placing a drop of medium beside the cover slip and letting some of it seep under the cover slip.

Before you place the cover slip on the medium containing protozoans in a wet mount, place a small quantity of cotton fibers on the medium. These serve as obstacles, blocking the path of protozoans and thus localizing their activities.

Use special media (glycerin, methyl cellulose, or commercially prepared products) to slow protozoans. Because these media are thicker than water, the protozoans move more slowly through them.

Dissection Techniques

In this course there are two lab activities in which you will be dissecting specimens. Before you do a dissection, make sure that you read the directions. Your teacher may also have you practice with a virtual dissection. Keep in mind that once you make a cut, you can't undo it!

Make sure that you have identified the correct structures by comparing them to drawings before you cut. Handle the specimens delicately. Preserved structures often tear and break easily.

If you will be finishing your dissection on a following lab day, you will need to label and preserve your specimen. Put your name and lab hour on a plastic bag with a permanent marker. Wrap the organism in a wet paper towel and put it in the bag. Gently squeeze out most of the air and tie or zip the bag closed. Organisms wrapped in this manner may be kept for a few days with good results.

Graphing Techniques

Constructing Graphs

When data is recorded in tables, it is difficult to see the relationship that exists between sets of numbers. To make trends and patterns easier to see, you will often put your data on a graph.

In experiments that search for a cause-effect relationship between two variables, you will cause one variable (the independent variable) to change and observe the effect on the second (the dependent variable). For example, if you were to investigate how the level of glucose in the blood changes with time after a meal, time would be the independent variable and blood glucose level would be the dependent variable. Traditionally, the independent variable is plotted on the x-axis of the graph and the dependent variable is plotted on the y-axis.

As you construct a graph, choose appropriate scales. Do not make the graph so small that the data cannot be clearly seen or so large that the graph will not fit on a single sheet of paper. Pick scales that include the entire range of each variable. Keep in mind that the scales on each axis do not have to be the same. For instance, the scale on the x-axis might be 1 h for every line, while the scale on the y-axis could be 50 mg/100 mL for every line. Also, your scale should be easy to subdivide. Sub-divisions of 1, 2, 5, and 10 units are the most convenient. Once you have settled on a scale, label the increments on the x- and y-axes. It is not necessary to label every increment. Labeling every second, fifth, or tenth increment will often make your axes look less cluttered.

When you have decided which variable will be plotted on which axis, include a label of each quantity and the units used to measure it on the appropriate axis.

Blood Glucose Level vs. Time

HOURS AFTER EATING (h)	BLOOD GLUCOSE (mg/100 mL)
1	140
2	175
3	120
4	110
5	80
6	75

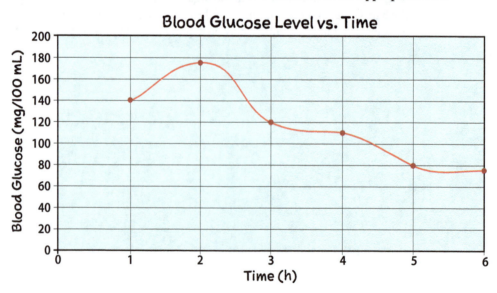

The title of the graph should state the variables (e.g., "Blood Glucose Level vs. Time") and be printed at the top of the graph. If more than one line will be sketched on the same graph, include a legend that identifies each line. Plot each of your data points by making small dots. Follow the specific lab guidelines for the proper way to handle these data points. You will usually draw a smooth curve through the data points. The graph on the facing page illustrates these techniques.

In some cases you will want to draw a straight line even though your data points do not fall precisely in a line. If this occurs, draw a line that shows the general relationship. Be sure to make the line go through the average values of the plotted points. In the first graph at right, the line is incorrect because it lies above the cluster of points near the bottom of the graph and below the cluster of points at the top. The second graph shows the correct method of fitting a straight line to a series of points.

Interpreting Graphs

The shape of a graph tells much about the relationship between the variables. When data appears to be arranged in a straight line, the x- and y-variables are related in a way that can be expressed as a linear equation. If this line goes through the origin, the linear equation will express the x-variable as a multiple of the y-variable (a direct variation). A positive slope means that the y-variable increases with the x-variable; a negative slope indicates that the y-variable decreases as the x-variable increases.

Data points that curve up (or down) from left to right indicate that data may be best modeled by some nonlinear equation. The equation relating the two variables may contain an exponent. Other possible relationships are inverse, exponential, or logarithmic; some can be modeled by a polynomial.

Graphs can be used to predict additional data points that have not been experimentally determined. Predicting points between data points on the basis of the graphed line is called *interpolation*. From the graph of blood glucose levels on the previous page it is reasonable to assume that the blood glucose level would be near 150 mg/100 mL at 2.5 h. Predicting values past the data points by extending the graphed line in either direction is called *extrapolation*. The graph of blood glucose level indicates that it could fall to 60 mg/100 mL at 7 h. This extrapolation is reasonable. The further from the data points we attempt to extrapolate, the less reasonable our predictions are likely to be. For example, if we tried to use the data provided to extrapolate out to 24 h, we would not make a reasonable prediction.

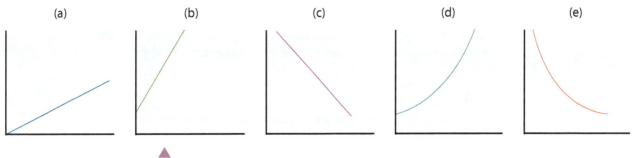

| (a) | (b) | (c) | (d) | (e) |

The graphs in this lab manual can have different shapes, showing linear relationships (a, b, c) or exponential relationships (d, e).

APPENDIX F

Writing Formal Lab Reports

As you have learned in your study of science, communication is an important part of the scientific process. Part of what scientists do is share what they have learned with other scientists. One example of this is the writing of formal lab reports. In a lab report a scientist presents a problem that was investigated, describes how the investigation was conducted, and summarizes what was learned during the investigation. Many college science courses require students to present all their laboratory work in formal lab reports.

To better prepare you for college, your teacher may choose to have you submit your lab assignments in formal lab reports rather than use the fill-in-the-blank format found in this lab manual. You may be more familiar with the latter form, but there really is little to be afraid of about writing formal reports. They may be new to you, and the word "formal" may sound alarming, but with practice such reporting can become second nature.

There really isn't a standardized way of writing and compiling a formal lab report. Your teacher may require you to do all your reporting in a bound journal, or you may be asked to submit individual typed reports. Regardless of which method your teacher prefers, there are some properties of lab reports that teachers generally agree upon. Let's look at the style and mechanics of a formal lab report and the type of content that is normally included.

Style and Mechanics

The first thing to keep in mind about a formal lab report is that it is a type of formal writing. That means that a formal lab report is like the final draft of an essay or research paper. It should have a neat, orderly appearance. To that end, formal lab reports normally have the following elements of style and mechanics.

» Unbound lab reports should be typed in black ink. Reports in a journal should be neatly written.

» Lab reports should be written in complete sentences and be checked for proper spelling, punctuation, and grammar.

» Formal lab reports should be free of erasures, whiteout, or scratched-out words.

» Since objectivity is a key aspect of the scientific process, lab reports usually avoid first-person narrative and instead use an impersonal third-person form.

Example:

INFORMAL: "I learned that there aren't any juvenile Chinook salmon upstream of Misty Falls."

FORMAL: "The survey indicated an absence of juvenile Chinook salmon upstream from Misty Falls."

Format

Scientists (and science teachers) may disagree on the exact number of components in a formal lab report, but there is a general consensus on what is presented and the order in which it is presented. Keep in mind as you write that a formal lab report answers three basic questions.

> » What problem was investigated?
> » What procedure was used during the investigation?
> » What was learned from the investigation?

To answer those questions, a formal lab report generally includes the elements shown below.

> » title
> » synopsis
> » introduction
> » list of any materials used and a description of the procedure
> » data yielded by the procedure
> » analysis of the data and a discussion of the findings
> » sources

Let's take a closer look at each of these.

TITLE

A formal lab report title should succinctly describe what the lab activity was about. It should be matter-of-fact rather than creative. Consider the following possible titles for a report on a Chinook salmon survey.

Example:

INFORMAL: Here, Fishy, Fishy!

FORMAL: The Distribution of Chinook Salmon within Boulder Creek

It's OK for a formal lab report title to sound dry—save the creative writing for your language arts class! Imagine yourself as another scientist who is reading your report as part of research and wants to quickly know what your report is about without having to read deep into the main body of the text.

SYNOPSIS

A synopsis is a very brief summary of the entire lab report, that is, two or three sentences. If the title of your report catches the reader's attention, the synopsis should then provide just enough detail to decide whether to read further.

Example (based on a presence/absence survey for juvenile Chinook salmon):

Three-pass electrofishing was used to sample five reaches within Boulder Creek to evaluate the distribution of Chinook salmon (*Oncorhynchus tshawytscha*). Juvenile Chinook salmon were found at only one sampling site, downstream from Misty Falls, suggesting that the falls act as a complete barrier to the upstream migration of Chinook salmon.

APPENDIX F

INTRODUCTION

Your introduction should expand on your synopsis by providing more details about the experiment. An introduction should include the elements listed below.

> » a description of the purpose of the experiment

> » the hypothesis that will be tested

> » a summary of the methods that will be used to test the hypothesis (Be brief. A more detailed description should be reserved for the Materials and Methods section that follows.)

> » a brief description of the results of the procedure

> » a summary of the conclusions that you have drawn on the basis of the results

MATERIALS AND METHODS

The goal of this section of your lab report is to describe your investigation in enough detail to allow another person to assess its validity or even replicate it. That is why there are two parts to this section: materials and methods. You should first list your materials, including names and quantities where appropriate (e.g., for chemicals). Your list should be just that—a list. Think about how you normally see such lists in this lab manual.

Next, describe your methods. You do not need to give instructions for how to do common tasks, such as using an electronic balance. Nor do you need to explain how to do common procedures if your audience is expected to be familiar with them. For example, in the population-sampling example used above it would not be necessary to explain how electrofishing is done; your audience probably already knows how to do this. What your audience would need to know in order to replicate your experiment is the type of electrofishing gear that you used along with the particular sampling methodology that was followed. As you write this section, ask yourself whether someone reading your descriptions would be able to replicate your experiment exactly. If the answer is no, then you need to be more specific. It is often helpful to include a diagram of any experimental setup used.

DATA

Your data section should be reserved for raw data only. This is where you report the results of your tests. Your data may be qualitative (i.e., based on observations) or quantitative (i.e., based on measurements). Often, a good way to report quantitative data is to use a data table. The format of data tables will vary depending on the data that you collected. An excellent technique, especially if you are using spreadsheets, is to record the units of measure in the column header and record only numerical data in the cells. This also allows you to use the computational features of the software. All quantitative data should be reported with the appropriate number of significant figures to indicate the precision of the measurements.

ANALYSIS AND DISCUSSION

» This is where the fun part begins! Now you get to interpret the data that you presented in the previous section. First, describe how the data was analyzed and show any formulas or equations that were used to do so. Sample calculations should be included. Calculated values should be reported with the proper number of significant figures. Descriptions of the results should include appropriate statistical parameters, such as the mean, median, and range of values. Where appropriate, use graphs to show trends in the data. All graphs should be properly scaled and include titles and axis labels with units.

» Once you have presented the analysis of your data, you may then address the all-important question: What does it mean? Think back to your stated hypothesis—did your data and analysis support your hypothesis? Were your results in line with what you expected? How did your results compare with those of others in your class? Be sure to support any claims you make with the relevant data and analyses. Your reader—your teacher—expects to see logical arguments that are consistent with your experimental results.

» Your discussion needs to be frank and honest. Don't try to make your data say something that it isn't actually saying. If your data does not support your hypothesis, then say so! Don't say that it does just because you think that might be the answer that your teacher is expecting. An important aspect of nurturing science skills is learning how to discuss inconclusive or contrary data. Examine what may have gone wrong during the experiment, make suggestions for improving future investigations, or propose a modified hypothesis for future testing.

» It is also appropriate during your discussion to suggest possible practical applications of your work or to put forward additional related questions that might be investigated in the future. In doing so, you might be laying the groundwork for a future scientist. By this you can see how the process of science has links both to those who have investigated a question in the past and to those who may investigate it in the future.

SOURCES CITED

» Much like any formal research paper, a lab report should cite sources within the body of the report and include a bibliography at the end. There are different formats for doing this. Your teacher will advise you on which format to use.

PHOTO CREDITS

Key: (t) top; (c) center; (b) bottom; (l) left; (r) right

COVER

Tim Platt/Stone via Getty Images

FRONT MATTER

i Eric Isselee/Shutterstock.com; **iii** "Robert Hooke, Micrographia, detail: microscope"/Wellcome Images/Wikimedia Commons /modified/CC By 4.0; **iv** bonchan/iStock/Getty Images Plus via Getty Images; **v** GMVozd/E+ via Getty Images; **vit** GL Archive /Alamy Stock Photo; **vib** Olha Rohulya/Shutterstock.com

CHAPTER 1

1 Motortion Films/Shutterstock.com; **7** "Portrait of Anthonie van Leeuwenhoek (1632-1723)" by Jan Verkolje/Wikimedia Commons /Public Domain; **8** 3DMI/Shutterstock.com; **10**l, r Rattiya Thongdumhyu/Shutterstock.com

CHAPTER 2

15 michaeljung/Shutterstock.com

CHAPTER 3

19t Rostislav Stefanek/Shutterstock.com; **19b** Morphart Creation /Shutterstock.com; **23** Vicki Jauron, Babylon and Beyond Photography/Moment via Getty Images; **25** dugdax/Shutterstock .com

CHAPTER 4

27t Mark A Paulda/Moment via Getty Images; **27b** © iStock.com /JulianneGentry; **33** Phynart Studio/E+ via Getty Images; **34** ANN PATCHANAN/Shutterstock.com; **37** (all) Adobe Stock/Calvin

CHAPTER 5

39 "Robert Hooke, Micrographia, detail: microscope"/Wellcome Images/Wikimedia Commons/modified/CC BY 4.0

CHAPTER 6

49t Brian A Jackson/Shutterstock.com; **49b** (art reference) © iStock .com/Bill Oxford; **55** isak55/Shutterstock.com; **56** Luis Line /Shutterstock.com; **57** "Phenol red pH 6,0 - 8,0" by Max schwalbe /Wikimedia Commons/modified/CC By-SA 4.0; **58** koto_feja /iStock/Getty Images Plus via Getty Images

CHAPTER 7

59 Brent Hofacker/Shutterstock.com; **65** Photo 12/Alamy Stock Photo

CHAPTER 8

69 Kateryna Kon/Shutterstock.com; **75t** duncan1890/DigitalVision Vectors via Getty Images; **75b** Photoongraphy/Shutterstock.com; **77** Barbara Rich/Moment via Getty Images; **80** Jurgis Mankauskas /Shutterstock.com; **81t** LittleMiss/Shutterstock.com; **81ct, b** JIANG HONGYAN/Shutterstock.com; **81cb, 83** Grant Heilman Photography /Alamy Stock Photo; **82t** GoodFocused/Shutterstock.com; **82b** ARTSILENSE/Shutterstock.com

CHAPTER 9

85l Eric Isselee/Shutterstock.com; **85r** pandapaw/Shutterstock.com; **88t** Jim West/Alamy Stock Photo; **88b** "Seedbank" by R. C. Johnson /Wikimedia Commons/CC By 2.5; **91** D-Keine/E+ via Getty Images

CHAPTER 10

97 Chronicle/Alamy Stock Photo; **98** Edward Westmacott /Shutterstock.com; **99** The Natural History Museum/Alamy Stock Photo; **100** "Charles Darwin 1865 signature"/Wikimedia Commons /Public Domain; **103t** Jean-Philippe Offord/Alamy Stock Photo; **103b** © Dr. Seth Shostak/Science Source; **104** kenkistler /Shutterstock.com; **106** Jeff Topping/Stringer/Getty Images Sport via Getty Images

CHAPTER 11

109t vkbhat/iStock/Getty Images Plus via Getty Images; **109b** "Meadow deathcamas - Zigadenus venenosus" by Neal Herbert /National Park Service/Wikimedia Commons/Public Domain; **112tl** "Channel catfish1" by Ellen Edmonson and Hugh Chrisp/Wikimedia Commons/Public Domain; **112tr** "Bluespotted Sunfish" by Ellen Edmonson and Hugh Chrisp/Wikimedia Commons/Public Domain; **112bl** "Rainbow Trout"/Wikimedia Commons/Public Domain; **112br** "Bowfin" by Ellen Edmonson and Hugh Chrisp/Wikimedia Commons/Public Domain; **113tl** Raver Duane, U.S. Fish and Wildlife Service/Public Domain Images; **113tr** "Moxostoma anisurum" by Ellen Edmonson and Hugh Chrisp/Wikimedia Commons/Public Domain; **113bl** "Bonneville Cutthroat trout" /UDWR/Flickr/Wikimedia Commons/CC By 2.0; **113br** "Muskellunge USFWS" by Knepp, Timothy - U.S. Fish and Wildlife Service/Wikimedia Commons/Public Domain; **114tl** "Lepomis auritus" by Duane Raver/Wikimedia Commons/Public Domain; **114tr** "Lake whitefish1" by Ellen Edmonson and Hugh Chrisp /Wikimedia Commons/Public Domain; **114bl** Raver Duane, U.S. Fish and Wildlife Service/Public Domain Images; **114br** "Longnose gar" by Ellen Edmonson and Hugh Chrisp/Wikimedia Commons/ Public Domain; **115tl** "Micropterus dolomieu smallmouth bass fish" by Raver Duane, U.S. Fish and Wildlife Service/Wikimedia Commons/Public Domain; **115tr** Raver Duane, U.S. Fish and Wildlife Service/Public Domain Images; **115bl** "Pumpkinseed fish lepomis gibbosus" by Raver Duane, U.S. Fish and Wildlife Service /Wikimedia Commons/Public Domain; **115br** "Black bullhead" by Ellen Edmonson and Hugh Chrisp/Wikimedia Commons/Public Domain; **116tl** "Esox lucius1" by Timothy Knepp/Wikimedia Commons/Public Domain; **116tr** "Black Crappie" by Ellen Edmonson and Hugh Chrisp/Wikimedia Commons/Public Domain; **116bl** "The Shad (Clupea Sapidissima)" by Shermon Foote Denton /Wikimedia Commons/Public Domain; **116br** "Largemouth bass fish art work micropterus salmoides" by Raver Duane, U.S. Fish and Wildlife Service/Wikimedia Commons/Public Domain; **117tl** Raver Duane, U.S. Fish and Wildlife Service/Public Domain Images; **117tr** Raver Duane, U.S. Fish and Wildlife Service/Public Domain Images; **117bl** "Yellow perch fish perca flavescens" by Raver Duane, U.S. Fish and Wildlife Service/Wikimedia Commons/Public Domain; **117br** "Brindled madtom" by Ellen Edmonson and Hugh Chrisp/Wikimedia Commons/Public Domain; **118tl** "White perch morone chrysops" by Raver Duane/U.S. Fish and Wildlife Service /Wikimedia Commons/Public Domain; **118tr** "White crappie pomoxis annularis" by Raver Duane, U.S. Fish and Wildlife Service /Wikimedia Commons/Public Domain; **118bl** "Brook trout" by Raver, Duane/Wikimedia Commons/Public Domain; **118br** "Acipenser oxyrhynchus" by Raver Duane/U.S. Fish and Wildlife Service/Wikimedia Commons/Public Domain; **121** "Whirling disease pathology"/NOAA/Wikimedia Commons/Public Domain; **122** "Carl von Linné" by Alexander Roslin/Wikimedia Commons /Public Domain/Modified

PERIODIC TABLE OF THE ELEMENTS

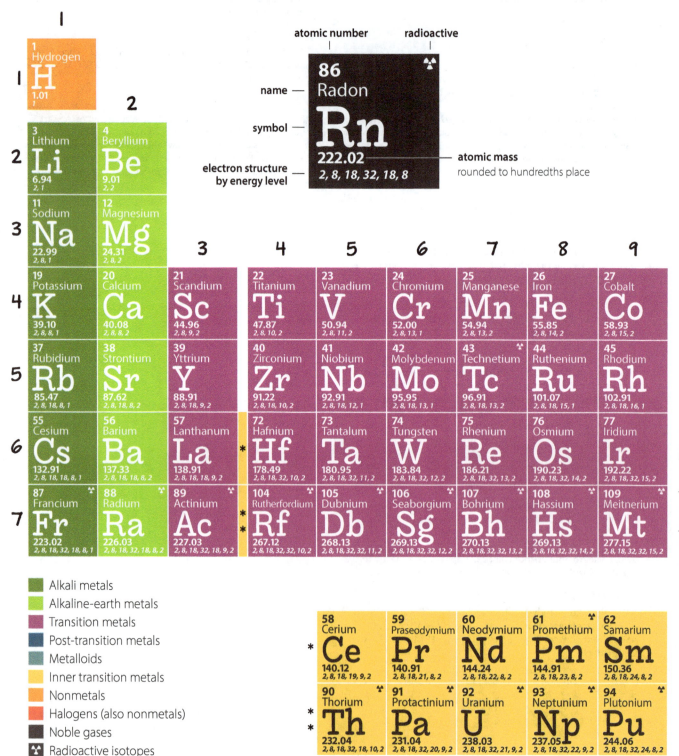

atomic number

radioactive

name —

86
Radon

symbol —

Rn

222.02

electron structure
by energy level

2, 8, 18, 32, 18, 8

atomic mass
rounded to hundredths place

1

1
1 Hydrogen **H** 1.01 *1*

2

3 Lithium **Li** 6.94 *2, 1*	4 Beryllium **Be** 9.01 *2, 2*
11 Sodium **Na** 22.99 *2, 8, 1*	12 Magnesium **Mg** 24.31 *2, 8, 2*

3 **4** **5** **6** **7** **8** **9**

19 Potassium **K** 39.10 *2, 8, 8, 1*	20 Calcium **Ca** 40.08 *2, 8, 8, 2*	21 Scandium **Sc** 44.96 *2, 8, 9, 2*	22 Titanium **Ti** 47.87 *2, 8, 10, 2*	23 Vanadium **V** 50.94 *2, 8, 11, 2*	24 Chromium **Cr** 52.00 *2, 8, 13, 1*	25 Manganese **Mn** 54.94 *2, 8, 13, 2*	26 Iron **Fe** 55.85 *2, 8, 14, 2*	27 Cobalt **Co** 58.93 *2, 8, 15, 2*
37 Rubidium **Rb** 85.47 *2, 8, 18, 8, 1*	38 Strontium **Sr** 87.62 *2, 8, 18, 8, 2*	39 Yttrium **Y** 88.91 *2, 8, 18, 9, 2*	40 Zirconium **Zr** 91.22 *2, 8, 18, 10, 2*	41 Niobium **Nb** 92.91 *2, 8, 18, 12, 1*	42 Molybdenum **Mo** 95.95 *2, 8, 18, 13, 1*	43 Technetium **Tc** 96.91 *2, 8, 18, 13, 2*	44 Ruthenium **Ru** 101.07 *2, 8, 18, 15, 1*	45 Rhodium **Rh** 102.91 *2, 8, 18, 16, 1*
55 Cesium **Cs** 132.91 *2, 8, 18, 18, 8, 1*	56 Barium **Ba** 137.33 *2, 8, 18, 18, 8, 2*	57 Lanthanum **La** 138.91 *2, 8, 18, 18, 9, 2*	72 Hafnium **Hf** 178.49 *2, 8, 18, 32, 10, 2*	73 Tantalum **Ta** 180.95 *2, 8, 18, 32, 11, 2*	74 Tungsten **W** 183.84 *2, 8, 18, 32, 12, 2*	75 Rhenium **Re** 186.21 *2, 8, 18, 32, 13, 2*	76 Osmium **Os** 190.23 *2, 8, 18, 32, 14, 2*	77 Iridium **Ir** 192.22 *2, 8, 18, 32, 15, 2*
87 Francium **Fr** 223.02 *2, 8, 18, 32, 18, 8, 1*	88 Radium **Ra** 226.03 *2, 8, 18, 32, 18, 8, 2*	89 Actinium **Ac** 227.03 *2, 8, 18, 32, 18, 9, 2*	104 Rutherfordium **Rf** 267.12 *2, 8, 18, 32, 32, 10, 2*	105 Dubnium **Db** 268.13 *2, 8, 18, 32, 32, 11, 2*	106 Seaborgium **Sg** 269.13 *2, 8, 18, 32, 32, 12, 2*	107 Bohrium **Bh** 270.13 *2, 8, 18, 32, 32, 13, 2*	108 Hassium **Hs** 269.13 *2, 8, 18, 32, 32, 14, 2*	109 Meitnerium **Mt** 277.15 *2, 8, 18, 32, 32, 15, 2*

* Lanthanum row
** Actinium row

Legend

- 🟩 Alkali metals
- 🟢 Alkaline-earth metals
- 🟪 Transition metals
- 🟦 Post-transition metals
- 🟦 Metalloids
- 🟨 Inner transition metals
- 🟧 Nonmetals
- 🟥 Halogens (also nonmetals)
- ⬛ Noble gases
- ☢ Radioactive isotopes

58 Cerium **Ce** 140.12 *2, 8, 18, 19, 9, 2*	59 Praseodymium **Pr** 140.91 *2, 8, 18, 21, 8, 2*	60 Neodymium **Nd** 144.24 *2, 8, 18, 22, 8, 2*	61 Promethium **Pm** 144.91 *2, 8, 18, 23, 8, 2*	62 Samarium **Sm** 150.36 *2, 8, 18, 24, 8, 2*
90 Thorium **Th** 232.04 *2, 8, 18, 32, 18, 10, 2*	91 Protactinium **Pa** 231.04 *2, 8, 18, 32, 20, 9, 2*	92 Uranium **U** 238.03 *2, 8, 18, 32, 21, 9, 2*	93 Neptunium **Np** 237.05 *2, 8, 18, 32, 22, 9, 2*	94 Plutonium **Pu** 244.06 *2, 8, 18, 32, 24, 8, 2*

18

13 **14** **15** **16** **17**

10 **11** **12**

							2 Helium **He** 4.00 2
			5 Boron **B** 10.81 2, 3	6 Carbon **C** 12.01 2, 4	7 Nitrogen **N** 14.01 2, 5	8 Oxygen **O** 16.00 2, 6	9 Fluorine **F** 19.00 2, 7
			13 Aluminum **Al** 26.98 2, 8, 3	14 Silicon **Si** 28.09 2, 8, 4	15 Phosphorus **P** 30.97 2, 8, 5	16 Sulfur **S** 32.06 2, 8, 6	17 Chlorine **Cl** 35.45 2, 8, 7

10
Neon
Ne
20.18
2, 8

18
Argon
Ar
39.95
2, 8, 8

28 Nickel **Ni** 58.69 2, 8, 16, 2	29 Copper **Cu** 63.55 2, 8, 18, 1	30 Zinc **Zn** 65.38 2, 8, 18, 2	31 Gallium **Ga** 69.72 2, 8, 18, 3	32 Germanium **Ge** 72.63 2, 8, 18, 4	33 Arsenic **As** 74.92 2, 8, 18, 5	34 Selenium **Se** 78.97 2, 8, 18, 6	35 Bromine **Br** 79.90 2, 8, 18, 7	36 Krypton **Kr** 83.80 2, 8, 18, 8
46 Palladium **Pd** 106.42 2, 8, 18, 18	47 Silver **Ag** 107.87 2, 8, 18, 1	48 Cadmium **Cd** 112.41 2, 8, 18, 2	49 Indium **In** 114.82 2, 8, 18, 3	50 Tin **Sn** 118.71 2, 8, 18, 4	51 Antimony **Sb** 121.76 2, 8, 18, 5	52 Tellurium **Te** 127.60 2, 8, 18, 6	53 Iodine **I** 126.90 2, 8, 18, 7	54 Xenon **Xe** 131.29 2, 8, 18, 8
78 Platinum **Pt** 195.08 2, 8, 18, 32, 17, 1	79 Gold **Au** 196.97 2, 8, 18, 32, 18, 1	80 Mercury **Hg** 200.59 2, 8, 18, 32, 18, 2	81 Thallium **Tl** 204.38 2, 8, 18, 32, 18, 3	82 Lead **Pb** 207.24 2, 8, 18, 32, 18, 4	83 Bismuth **Bi** 208.98 2, 8, 18, 32, 18, 5	84 ☢ Polonium **Po** 208.98 2, 8, 18, 32, 18, 6	85 ☢ Astatine **At** 209.99 2, 8, 18, 32, 18, 7	86 ☢ Radon **Rn** 222.02 2, 8, 18, 32, 18, 8
110 ☢ Darmstadtium **Ds** 282.17 2, 8, 18, 32, 32, 16, 2	111 ☢ Roentgenium **Rg** 282.17 2, 8, 18, 32, 32, 17, 2	112 ☢ Copernicium **Cn** 286.18 2, 8, 18, 32, 32, 18, 2	113 ☢ Nihonium **Nh** 286.18 2, 8, 18, 32, 32, 18, 3	114 ☢ Flerovium **Fl** 290.19 2, 8, 18, 32, 32, 18, 4	115 ☢ Moscovium **Mc** 290.20 2, 8, 18, 32, 32, 18, 5	116 ☢ Livermorium **Lv** 293.21 2, 8, 18, 32, 32, 18, 6	117 ☢ Tennessine **Ts** 294.21 2, 8, 18, 32, 32, 18, 7	118 ☢ Oganesson **Og** 295.22 2, 8, 18, 32, 32, 18, 8

63 Europium **Eu** 151.96 2, 8, 18, 25, 8, 2	64 Gadolinium **Gd** 157.25 2, 8, 18, 25, 9, 2	65 Terbium **Tb** 158.93 2, 8, 18, 27, 8, 2	66 Dysprosium **Dy** 162.50 2, 8, 18, 28, 8, 2	67 Holmium **Ho** 164.93 2, 8, 18, 29, 8, 2	68 Erbium **Er** 167.26 2, 8, 18, 30, 8, 2	69 Thulium **Tm** 168.93 2, 8, 18, 31, 8, 2	70 Ytterbium **Yb** 173.05 2, 8, 18, 32, 8, 2	71 Lutetium **Lu** 174.97 2, 8, 18, 32, 9, 2
95 ☢ Americium **Am** 243.06 2, 8, 18, 32, 25, 8, 2	96 ☢ Curium **Cm** 247.07 2, 8, 18, 32, 25, 9, 2	97 ☢ Berkelium **Bk** 247.07 2, 8, 18, 32, 27, 8, 2	98 ☢ Californium **Cf** 251.08 2, 8, 18, 32, 28, 8, 2	99 ☢ Einsteinium **Es** 252.08 2, 8, 18, 32, 29, 8, 2	100 ☢ Fermium **Fm** 257.10 2, 8, 18, 32, 30, 8, 2	101 ☢ Mendelevium **Md** 258.10 2, 8, 18, 32, 31, 8, 2	102 ☢ Nobelium **No** 259.10 2, 8, 18, 32, 32, 8, 2	103 ☢ Lawrencium **Lr** 262.11 2, 8, 18, 32, 32, 8, 3